Greening Your Home

About the Author

Clayton Bennett has more than 20 years of experience as a nonfiction writer. He wrote *Ortho's All About Decks*; *Ortho's All About Patios*; and *Ortho's All About Backyard Structures*; and edited *Ortho's All about Roofing and Siding Basics*. Bennett also wrote *A Portfolio of Home Entertainment Ideas* for Cowles Creative Publishing, revised the *Complete Guide to Floor Decor* for Black & Decker, and contributed to the Black & Decker books *Complete Guide to Home Plumbing*; *Porch, Patio & Deck Furnishings*; *Advanced Deck Building*; and *Complete Photo Guide to Home Repair*. He also writes technical and marketing materials for organizations with complex products and services.

Greening Your Home

Sustainable Options for
Every System in Your House

Clayton Bennett

New York   Chicago   San Francisco
Lisbon   London   Madrid   Mexico City
Milan   New Delhi   San Juan
Seoul   Singapore   Sydney   Toronto

The *McGraw·Hill* Companies

Cataloging-in-Publication Data is on file with the Library of Congress.

McGraw-Hill books are available at special quantity discounts to use as premiums and sales promotions, or for use in corporate training programs. To contact a special sales representative, please visit the Contact Us page at www.mhprofessional.com.

Greening Your Home

1 2 3 4 5 6 7 8 9 0   DOC/DOC   0 1 4 3 2 1 0 9 8

ISBN 978- 0-07-149909-5
MHID 0-07-149909-1

Sponsoring Editor Joy Bramble Oehlkers	**Proofreader** Divya Kapoor
Acquisitions Coordinator Rebecca Behrens	**Indexer** Deborah Morse-Kahn
Editorial Supervisor David E. Fogarty	**Production Supervisor** Richard C. Ruzycka
Project Manager Preeti Longia Sinha	**Composition** International Typesetting and Composition
Copy Editor Ragini Pandey	**Art Director, Cover** Jeff Weeks

Contents

Preface

For years, environmentally sound practices were associated with a counterculture. Resource conservation, energy independence, and so on were perceived as the rejection of mainstream values. Recently, though, more people have realized that they reflect the American characteristics of innovation and individualism. Green living is no longer a social expression of political ideas, but an economic expression of personal ones.

The thirty subjects covered in this book range from non-traditional materials to advanced technologies to age-old practices. Some are better suited for new construction, others for remodeling. Most can be adopted or adapted no matter where you live. What they all have in common is that they make better ecological sense than the materials and methods they replace.

Most of us take several factors into account when we decide how to build or remodel the places we live. Quality and cost are the most obvious, along with comfort, appearance, durability, and other things that can be harder to define. All the products and practices described in this book can help you improve your standard of living, and exercise greater self-reliance, now and in the future.

Many people helped bring this book together; a few deserve special mention. In addition to the manufacturers and distributors who provided technical details and photographs, I thank Joy Bramble Oehlkers at McGraw-Hill for her encouragement and guidance, contractors Howard Kroll and Wade Clarke for their practical expertise, and my family for everything else.

Clayton Bennett

Material Changes.

Benefits and Trade-offs

In some ways, replacement materials provide the easiest means to improve the environmental friendliness of your house. Traditional flooring surfaces, landscaping, foundation walls, insulation products, and other materials can be replaced with alternatives that reduce impact on the natural environment. Before you read about some of these materials and their benefits, however, remember to choose which benefits are the most important to you. No single product will meet every criterion on your wish list.

Criteria for Selection

Here are the characteristics most people have in mind when they look for building products that offer environmental advantages:

1. Environmental Impact: Structural Insulated Panels
2. Health Impact: Low-VOC Paints
3. Availability: Replacement Windows
4. Maintenance: Green Certified Doors
5. Durability: Glass Block
6. Sustainability: Fly Ash Concrete
7. Renewability: Bamboo and Cork Flooring
8. Recyclability: Plastic Plumbing Pipe
9. Simplicity: Red Cedar Siding
10. Efficiency: Recycled Insulation

Setting Your Priorities

Every choice you make in updating your living space will involve at least one of these characteristics. The trade-offs are sometimes obvious. For example, switching your heating fuel from natural gas to wood can improve renewability while increasing maintenance requirements. And using synthetic lumber made from recycled milk jugs helps keep materials out of the waste stream, but with generally higher processing and shipping costs.

Instead of trying to create a catalog of all current materials being promoted as environment-friendly, this section will provide 10 examples that meet a few of these characteristics—one from each of 10 types of material. From these illustrations, you can make better informed decisions about the properties you feel are the most important.

CHAPTER 1

Environmental Impact

Products that use fewer toxic chemicals or produce lower harmful emissions in manufacturing

Structural Insulated Panels

1.1 Getting More from Less

With so many people concerned about deforestation, you might not expect wood products to be the first things mentioned in this book. But some engineered wood products like these are made more efficiently. The result is high-quality new construction and remodeling with reduced overall consumption of natural resources.

According to APA, an industry group, wood products make up nearly half of the industrial raw materials manufactured in the United States, but consume only four percent of the energy needed to manufacture those materials. The forest products industry recognizes the importance of trees—not just after they are harvested, but while they are growing. For every ton of wood grown, a young forest produces 1.07 tons of oxygen and absorbs 1.47 tons of carbon dioxide. Recent figures show that, for every 100 trees harvested, 125 are planted.

1.2 Making a Sandwich

Structural insulated panels (SIPs) are ready-made sections for use in walls, roofs, or even floors. They are also known as foam core panels or sandwich panels. Several varieties are available, with slight differences in materials and applications. All SIPs feature a layer of foam

Photo courtesy of APA—The Engineered Wood Association

insulation sandwiched between and bonded to two sheets of wood, usually oriented strand board or plywood.

The foam insulation offers an effective barrier against heat, cold, and sound; the wood sheets provide strength and stability. The panels deliver high insulating and structural performance with light weight. These versatile panels can be used in place of traditional wall framing, roof decking, and subflooring. Some companies also make SIPs with an additional layer, offering interior gypsum wallboard or exterior siding already installed.

1.3 Choosing Ingredients

The foam in an SIP can be one of several types, including extruded polystyrene, urethane, and polyisocyanurate. The most commonly used foam is expanded polystyrene (EPS), which is relatively easy and inexpensive to make. All the foams used in SIPs are light and strong. They keep their shape over time, resist moisture, and do not contain formaldehyde.

Although SIPs depend in part on petroleum products, they provide considerable environmental benefits. This is a trade-off that requires accepting one loss to realize another gain.

1.4 Features and Advantages

The strength and rigidity of SIPs make them well suited for use in walls, both for their ability to support axial loads (tension or compression bearing on the long axis of a straight structural member) and their resistance to lateral forces, such as strong winds.

Along with strength, SIPs have impressive insulating qualities. Each panel functions as a single structural part. Because the plastic foam core separates the two wood faces, temperature differences between the outside and inside are less likely to be transmitted through connecting pieces—like traditional 2 × 4 studs. The foam creates a uniform barrier.

Depending on what type of foam is used, an SIP wall provides between R4 and R7 for each inch of thickness. That can translate to more than twice the R-value of a comparable stud wall. An average 2 × 4 stud wall with fiberglass or mineral wool insulation has R-values of 11 to 15, while an SIP of the same thickness provides R-values of 14 to 25.

In addition to higher R-values per inch of thickness, SIP walls reduce air infiltration. This means even greater benefits for climate control. Even when a research group had two houses built side by side to matching specifications and R-values, the one built with SIPs consumed about 15 percent less energy for heating. Construction using SIPs is virtually airtight.

Such a well-sealed envelope may not always be desirable, of course. A house with walls made of SIPs may have so little air infiltration that the house may need a fresh-air ventilation or heat-recovery system. As with concerns about airborne environmental hazards, though, this means the homeowner has more control over what goes through the house. And, with less air flowing through and around the walls, there is also less noise.

Through its Energy Star program, the U.S. Environmental Protection Agency recommends SIPs as energy-efficient building materials.

1.5 Using Structural Insulated Panels

Structural insulated panels can help you simplify a project, whether you're replacing walls, adding on, or building something entirely new. Each panel contains three components in one: structural members, insulation, and sheathing. While SIPs are made in all sizes, most are 8 feet high and anywhere from 4 to 24 feet wide. The SIPs made for residential construction can be cut at the jobsite and installed by hand with just two people.

Because they take the place of traditional framing members, SIPs must be made to provide at least the same amount of structural support. The panels are created for specific uses, based on the building plans. These plans, some of which are sold by companies that manufacture SIPs, provide the basis for how the panels are made. The parts are not necessarily interchangeable.

To make sure the panels with cutouts for doors or windows are sufficiently strong, use the panels from a building package that are intended for those purposes. If you cut your own panels on site, make sure you follow the manufacturer's recommendations. When you create an opening that is more than 4 feet wide, you will need to add a header. For all openings, plan to install 2× wood framing members along the cut edges. This will also provide a nailing surface and protect the foam insulation.

Manufacturers also provide wiring chases within the panels, so you can run electrical, phone, cable, and network cables without compromising the panel's strength or insulating properties.

The fire resistance of any structural insulating panel will depend on the wood products and type of foam used. Most national and local building codes provide for SIPs. Some regulations require a thermal barrier, such as $1/2$-inch gypsum board, on the interior surface.

This is one area for following the directions exactly. Your goal is to build a structure that is safe and strong. It must meet or exceed the requirements of your local building codes to be financed, insured, inspected, or sold in the future.

1.6   Putting It Together

Building with SIPS should be easy and fast. Thanks to their consistent nailing surfaces, you can join wall sections securely. You can also fasten fixtures, siding, cabinets, and so on anywhere you like. Just be sure to buy fasteners intended for use with SIPs. The consistent thickness and density of SIPs also make them less vulnerable to damage, whether it's accidental or intentional.

Although different manufacturers have their own methods for joining individual panels, two approaches using splines are the most common. In the first method, 2× wood splines are attached between the panels; they function like studs if they are also fastened to bottom plates and headers. In the second method, the splines are thinner, wider boards that fit in grooves along the edges of each panel. The splines are held in place with adhesives or fasteners, and the seams are sealed with caulk to minimize air infiltration.

Resources

For more information on SIPs, visit the Structural Insulated Panel Association web site at *www.sips.org*.

CHAPTER 2

Health Impact

Materials to replace substances that are unhealthy to people, animals, or plants

Low-VOC Paints

2.1   Covering the Walls

Painting is perhaps the most popular remodeling project of all. It uses common skills and equipment, and creates a large change for a relatively small investment of time and money. Until recently though, most paints still contained harmful substances known as volatile organic compounds (VOCs).

This is not the first time paint formulas have been changed for the sake of public health. Until the 1970s, many paints contained pigments made with lead, a heavy metal. It helped the paint dry more quickly and last longer, but its use was linked to serious health problems, especially in small children. Lead paint was prohibited from indoor use in the United States in 1978. The legal limit for lead content in paint is now one-half of one percent by weight.

If you are concerned that your house may still contain lead-based paint, buy a test kit at any hardware store or building supply center. If you find high levels of lead, call your local health authority or pollution control agency for advice on eliminating it.

2.2   Changing the Formula

Although paint manufacturers got the lead out years ago, they still use VOCs. Most are petroleum-based solvents, which help the paint flow smoothly and dry quickly. The gases released by these paints contain acetone, ethylene, methylene chloride, toluene, and xylene.

Photo courtesy of American Formulating & Manufacturing

These are listed as sources of indoor air pollution—and some paints can take years to finish "offgassing."

Small children, pregnant women, and people with allergies or asthma are especially sensitive to strong odors, including those given off by VOCs. But anyone can sense the effects that VOCs have on their health. Exposure to VOCs can cause headaches, dizziness, nausea, fatigue, and eye irritation. Over the long term, this kind of exposure can lead to damage in the liver, kidneys, and central nervous system.

2.3   Clearing the Air

To turn a painting project into an exercise in environmental improvement, look for finishes with lower levels of volatile organic compounds. These products have water-based mediums instead of petroleum-based solvents, and generally contain little or no formaldehyde.

In order to claim low VOC content, the manufacturer must show their content in consistent terms. The current limits allowed by the U.S. Environmental Protection Agency (EPA) are 200 grams of VOCs per liter of paint or stain, and 300 grams of VOCs per liter of varnish. Those numbers are higher than the actual levels for many products, which may have as little as 50 grams of VOCs per liter. Paints with the lowest certified levels of volatile organic compounds carry the Green Seal (GS11) mark.

2.4   Comparing Products

No two kinds of conventional paint are quite the same, but they have a lot more in common than any two paints that offer environmental benefits. The reason is simple: major paint manufacturers have competed on price and performance, which has led to most of them creating very similar products. When the goals of health and safety are just as important, more options become available.

Materials used in these new paints range from chalk and talcum to natural latex to clay to mineral dyes to milk casein to beeswax to plant oils and resins. While some people may be allergic or sensitive to even these products, they have far fewer adverse health effects. Water-based paints have generally lower odors, and some paints that use natural oils for a medium are actually pleasant to smell.

Manufacturers include (in alphabetical order) American Formulating and Manufacturing (AFM), Aglaia, Allied PhotoChemical, American Pride, Anna Sova, Auro, Benjamin Moore, Best Paint, BioShield, Cloverdale, Devoe, Frazee, Green Planet, ICI, Kelly-Moore, Livos, MAB Paints, Miller Paint, Olympic, PPG, Real Milk Paint, Sherwin Williams, Silacote, Vista Paint, Weather-Bos, and Yolo Colorhouse.

Some of those names will be familiar; most well-known brands now offer environment-friendly options. Other names may be new to you, but give them a look as well. If you want to support companies that make safer, healthier products, you may prefer to buy from one that is dedicated to the same goals.

Large building centers will stock at least a few paints with low VOC content. Independent paint stores are likely to have a greater variety of brands and range of quality, if perhaps higher prices. In some cities, specialty retailers carry environment-friendly building products of all kinds. These may be the best places to explore your options before you buy.

2.5   Other Surface Treatments

Paint is not the only surface treatment you apply with a brush, roller, or sprayer. Other coating and sealing products include

- Wallboard primers
- Wood and metal primers

- Wood stains and sealers
- Varnish and lacquer replacements
- Masonry and grout sealers
- Paint strippers

More ways to save

Whatever kind of paint you decide to use, here are some additional techniques for improving health and safety, conserving materials, and saving money.

- Choose a surface treatment first for how well it is suited to your project, and second for its other properties.
- Check VOC content. Look for paints that have 50 or fewer grams per liter.
- Watch for labels from health and safety agencies. They usually contain information about the risks of using that product.
- Compare the ratio of pigment (coloring) to medium (liquid). Paints with more pigment tend to cover better, reducing the amount you need to use.
- Make careful estimates of the amount you need. Start with the general rule that one gallon of paint covers about 400 square feet.
- Test older paint. If the existing paint was applied before 1980, get a test kit and check for lead content. Ask your local health authorities for advice before removing lead paint.
- Keep the air moving. While you want to avoid dust and dirt from getting on freshly painted surfaces, you do want the air to circulate.
- Keep brushes and rollers fresh. If you plan to use them again soon, wrap them in airtight plastic instead of washing them every time. (Don't try this with lacquer or varnish.)
- When possible, use turpentine; it's made from natural sources, and works well for cleaning oil paint off brushes.
- Make the most of your solvents. Let any paint sediment settle, then pour the clear solvent into a new container and use it again.
- Dispose of leftover paint responsibly. Many cities have hazardous waste drop-off sites, where you can leave unused paint for others to take. This helps others save money as well.

Resources

For more information on low-VOC paints and finishes, visit *www.builditgreen.org*.

CHAPTER 3

Availability

Items in local supply that require less effort and expense to get

Replacement Windows

3.1 Setting Priorities

Home improvement projects can serve several purposes. You might want to fix something that has been a problem, or make part of your house more comfortable, or reduce the time and money you spend on maintenance, or increase the resale value of your house, or save on heating and cooling expenses over the long run. Energy-efficient windows help you achieve all those goals at once.

3.2 Seeing the Benefits

No matter how large your house is, or how small, windows account for an important part of your heating and cooling costs. Because most window surfaces have lower insulating values than the surrounding walls, and because the frames around windows are often poorly sealed, you can lose as much energy from all your closed windows as if one was left wide open.

When only a few companies made windows with multiple panes, low-emissivity coatings, exterior cladding, and so on, shipping could make up half the price of a window. They are large and heavy, and many have to be custom-built to fit a particular opening. But because energy-efficient windows benefit homeowners in every climate, they have become very popular for both new construction and remodeling projects. Manufacturers and distributors now offer them across the country.

Photo courtesy of Kolbe & Kolbe Millwork

Just about any existing single-pane window can be replaced with an energy-efficient model. The price of replacement windows may seem high, but they bring greater savings for a longer time than other improvements of similar cost. They also make your house more comfortable in all seasons. So, while most updates to your house count as investments, this one brings real returns as soon as you finish.

Few other home improvements yield as many benefits as energy-efficient replacement windows. In addition to savings on heating and cooling, and noticeably greater indoor comfort, new windows add to the resale value of a house. In fact, the increase in resale price is often greater than the price of installing the new windows.

3.3   Common Features

Every major window manufacturer offers energy-efficient options. Each has its own set of design characteristics, but most models share the following properties:

- *Insulating frame materials*: Newer wood composites and synthetic lumber provide strong frames with better insulating qualities than traditional wood frames.

- *Multiple panes*: Two layers of glass will reduce heat transfer, and three can provide both added thermal insulation and soundproofing.

- *Thermal spacing*: A gap between the panes further improves insulation. The material used as a spacer can also help reduce heat transfer and condensation.

- *Gas fills*: Inert gases like argon or krypton between the panes will provide better R-values than regular air.

- *Low-emissivity glass*: Outer surfaces treated with low-emissivity coatings allow light to pass through, but reflect infrared and ultraviolet light. This keeps heat out in the summer and during winter, and protects your furniture and artwork from sunlight.

Some replacement windows also feature exterior cladding—an extra layer of vinyl or aluminum that offers protection against the elements. Clad windows provide slightly higher R-values than windows without cladding. Even better, they might never need to be painted.

3.4   Estimating the Costs

With many projects intended to reduce energy consumption, the cost-benefit ratio is relatively small. Weather stripping and caulk may be inexpensive, but they take lots of time and labor to install. Because the gains are hard to see, the effort is harder to make. Replacing an entire window requires much more work than sealing the edges of an existing one, but the benefits are immediate and obvious.

According to research figure from Lawrence Berkeley National Laboratory, if you install energy-efficient windows in San Diego, which has a warm and stable climate, you can save nearly 4 million Btu per year, which translates to $65 in current heating costs. That makes the payback time fairly long—unless energy prices go up. In Great Falls, Montana, where the temperature and humidity change dramatically over the course of a year, an energy-efficient window can bring savings of roughly 62 million Btu, or $695 per year. If energy prices rise, that figure will go higher as well. The payback time would be almost immediate.

3.5   Comparing Products

Building supply centers, lumberyards, and specialty stores can all stock energy-efficient windows, but comparison shopping can be difficult. Few retailers carry more than one brand, which means you have to visit several places to check one set of windows against another. And while the common features listed above apply to most replacement windows, manufacturers try to set themselves apart by giving similar products different names, and emphasizing the qualities their windows offer while minimizing any weak points.

To see how different windows compare, use the current standard for technical information—the Energy Star rating system from the U.S. Environmental Protection Agency (EPA). It will help you check the measurable differences in performance among all products that qualify for Energy Star labeling. These tests and measurements are set by a nonprofit industry group called the National Fenestration Rating Council (NFRC).

3.6   Measuring Performance

Any window that qualifies for the Energy Star designation must include information on the following characteristics:

- *U-Factor:* This tells you the rate of heat transfer, which indicates the level of insulation provided. Unlike R-values, where higher numbers mean better performance, you will want to find windows with the lowest U-Factor ratings.

- *Solar heat gain coefficient:* This is a measure of how much heat from sunlight gets through the window and into your house. Values range between zero and one; again, lower numbers mean better performance.

Those two values determine the Energy Star rating for a window. An NFRC label also provides information on other performance characteristics:

- *Visible transmittance:* Similar to solar heat gain coefficient, this is a measure of the light each window allows to pass through. These values also range between zero and one. The difference is that you generally want less heat gain but more light transmittance.

- *Condensation resistance:* In locations where differences in humidity cause excessive condensation, this rating becomes very important. Ranging from 0 to 100, this figure indicates a window's resistance to condensation; higher ratings mean greater resistance to water buildup.

- *Air leakage(AL):* This figure shows how much air gets past the window in a set period: cubic feet of air per square foot of window per minute (cfm/ft^2). Lower values mean less air leakage. Most building codes require an AL value of 0.3 cfm/ft^2.

You don't need to become an expert on heat transfer to make an informed buying decision. Just remember that lower numbers are better for U-factor, solar heat gain coefficient, and air leakage; higher numbers are better for visible transmittance and condensation resistance.

Resources

For now, the best single source of information on energy-efficient windows is the EPA. Its Energy Star Web site includes explanations of the different terms you will encounter, guides for regional climate requirements, lists of manufacturers, buying advice, and even tips on finding rebates or tax credits to help you reduce your expenses. Visit *www.energystar.gov* for more information. You can also learn more about NFRC labels at *www.nfrc.org.*

CHAPTER 4

Maintenance

Materials and products that require minimal upkeep

Green Certified Doors

4.1 Reducing Operating Costs

Other than windows, the main openings to your house are the doors. Installing a new door or replacing an existing one requires several kinds of skills—accurate measurement, careful alignment, balanced adjustment, secure fastening, and thorough weatherproofing. You will have to decide whether to hang a door yourself or have a contractor do it. Choosing the door, however, only requires some basic knowledge.

Reducing maintenance can be environmentally beneficial, especially if the materials used for maintenance create waste or contain VOCs. Doors used in commercial buildings are made to withstand heavy use for many years with little care; they are often made of steel and covered in a tough enamel finish. But that doesn't mean you need to use industrial products at home. Newer doors made of wood products can provide the same benefits while adding to the beauty, function, security, and insulation of your house.

4.2 Using Common Terms

One challenge of choosing a door is finding comparable information about more than one model or brand. Unlike appliances, most of which carry Energy Star ratings, doors can be rated on several criteria that are not always the same. The only doors that are required to have Energy Star ratings are those with windows. That's because the energy efficiency of the glass is measured and rated; the rest of the door isn't even considered.

iStockPhoto

For doors without windows, Energy Star will not help. Manufacturers promote the features that cast their products in the most favorable light, which makes more work for you. The National Association of Home Builders (NAHB) has a set of guidelines for applying green building principles in a construction project. These guidelines provide information that will help you determine which characteristics are the most important to you—but they won't appear on the products you consider. It's still up to you to weigh your choices.

To offer a system that contractors and consumers can use as a common reference, the U.S. Green Building Council introduced the Leadership in Energy and Environmental Design (LEED) Green Building Rating System. The LEED system began as a voluntary effort to create common standards for green building. The goal has been to share information and provide incentives for builders to adopt green practices rather than depend on government regulation to shape the industry.

According to its publications, "LEED is a collaborative initiative that actively involves all sectors of the home building industry including builders, home owners, product manufacturers, service providers, and affordable housing developers." While most of the work on LEED certification concentrates on new construction, the organization also offers the LEED for Homes Rating System. As more manufacturers and building products retailers adopt the LEED system, you will have more consistent terms and measurements to compare.

4.3   Adding Up the Points

One of the ways LEED measures environmental responsibility is by assigning points to each of several practices. The more green practices a manufacturer uses, the more points are given to the resulting product. Three of the practices measured by LEED are recycled content, regional availability, and rapid renewability.

- *Recycled content:* Using five percent postconsumer recycled content, or 10 percent combined postconsumer and postindustrial recycled content, earns a basic LEED point. Additional credit is given to products made with higher amounts of recycled material.

- *Regional availability:* Using at least 20 percent of materials and products that are grown or manufactured within 500 miles of the building site. This represents an effort to reduce transportation costs and return to a focus on the use of regional building supplies.

- *Rapid renewability:* Using natural materials that can be regenerated within 10 years; this includes products such as cork, which can be harvested every nine years without felling the tree, and bamboo, which is actually a kind of grass.

4.4   Comparing Features

Doors are among the simplest, oldest, and most common manufactured building products. For centuries, most of them were made from wood and finished with paint, stain, or sealer. Until the use of manmade materials in the doors and petroleum products in the finishes, most of them would meet today's definitions of environmental responsibility. Current green products may appear to be just a return to old ways of doing things, but there's much more to them.

Because the manufacturing process is similar for most kinds of doors, you can find green options in every design, both for interior and exterior uses. The terms you will encounter first are *stile and rail* and *flush*. Stile and rail refers to the separate pieces of material, usually wood, that make up a door: stiles are the vertical pieces, and rails the horizontal ones. Flush doors have a uniform surface and no visible connections.

You can find stile and rail doors in many designs. Within the stiles and rails are solid panels, usually with square corners but sometimes with contoured shapes, window glass (called "lights"), louvers or a combination of these things. Flush doors offer fewer options because the surface is usually all one material and finish. Doors of both kinds are also made with lower-quality surfaces for customers who plan to apply other finishes. These are called "paint grade" doors.

While you're evaluating product designs and properties, you may also want to consider additional features. For example, if you're remodeling or adding on to a historic house, look for manufacturers that offer period styles or custom millwork. If you're creating a home office or master suite, think about the benefits of acoustic insulation, which reduces the sound that passes through a door. And, depending on your requirements, think about the possible safety benefits of adding a fire-rated door. Bullet-resistant doors are also available—but if you need one, environmental responsibility may not be your top priority.

The millwork department of your local lumberyard or building supply store is the best place to start looking at the different kinds of doors available. It can also be the best place to order doors that are not regular stock items; the personal service and professional advice you can get from a millwork specialist can be worth far more than the money you save by ordering something sight unseen from far away.

Resources

For more information about the LEED system, visit *www.usgbc.org*.

CHAPTER 5

Durability

Products that last longer, requiring less frequent replacement

Glass Block

5.1 Making Sound Investments

Part window, part wall—glass block has properties that set it apart from all other building materials. While they are most often used in commercial and public buildings, glass blocks and panels can be used to create clear divisions between residential spaces. Suitable for non-structural interior and exterior applications, glass block can enhance or even define a kitchen, bathroom, laundry room, entryway, breeze-way, or other living area.

5.2 Controlling Light and Heat

The first environmental benefit of glass block is its transparency. By allowing natural light from the outside in, or between interior rooms, glass block reduces the need for artificial lighting. Natural daylight also helps some people keep their emotional balance, or regulate cycles of waking and sleep.

Unlike sheet glass, however, glass block maintains privacy. Depending on the surface pattern and overall thickness, a window or wall made with glass block can obscure vision without requiring other coverings, such as blinds or curtains. The styles that provide the greatest privacy, such as those with fibrous glass inserts or thick-faced block, still transmit at least 50 percent of the visible light that strikes them. With solid glass block, 80 percent of the light is transmitted. Flat sheet glass allows 90 percent of light through, but without the many advantages offered by glass block.

iStockPhoto

Each glass block is made of two halves, containing silica sand, soda ash, and limestone, and poured separately. The halves are combined while hot, and then placed in an oven for annealing—to cool slowly and increase in strength. This creates a partial vacuum within the block, which helps improve insulation and control condensation later on. An edge coating of polyvinyl butyral (PVB) resin allows for expansion and contraction, and helps mortar bond to the glass.

Compared to sheet glass, which has a heat transmission value (U-value) of 1.04, glass block provides far better insulation. Solid glass block has a U-value of 0.87, while standard glass block has a U-value of 0.51.

5.3   Other Environmental Benefits

Another benefit is impermeability. Because glass block is a masonry product, it keeps out all the elements. Correctly sealed, it can provide weather resistance equal or superior to any other material. Where air or water infiltration is a problem, glass block can help you control the indoor atmosphere. This becomes important when you or family members are sensitive to airborne dust and pollen, or when moisture control affects livability.

Insulation is another benefit of using glass block. The space between the two faces provides insulating value comparable to a double-pane window, without any drafts or leaking. This buffering space also minimizes condensation on the surface. Unless glass block is on an exterior wall facing direct exposure to sunlight, it brings both improved control over the interior environment and reduced heating and cooling costs.

One feature every homeowner wants to have, but no one wants to test, is resistance to fire. Glass block is available with fire ratings of 45, 60, or 90 minutes of resistance. Because glass is considered a window material instead of a wall material, though, this rating only affects how well glass block remains intact, and prevents the passage of smoke and fumes, during a fire. If you want to see the details, look for the standards ASTM E-2010 and NFPA 257.

One more thing: Glass block is made using simple processes and common materials, which translates to reduced environmental impact in manufacturing. Glass is made without the use or production of VOCs; it can also be broken down and recycled after use, helping reduce disposal work, landfill volumes, and replacement material costs. Its life cycle impact, one way of measuring environmental effects, is impressively low.

5.4   Exterior and Interior Uses

Most glass blocks made and sold today are either 3 or 4 inches nominal. The actual thicknesses are $3\frac{7}{8}$ inches, and $3\frac{1}{8}$ inches. Regular glass block, the thicker of the two, is intended for applications that require durability, insulation against temperature changes and sound, and resistance to fire. This thickness offers the greatest variety of surface patterns and block sizes. Thin glass block is usually meant for use in ready-made panels, such as sidelights for entryway doors. This glass block can only support its own weight, not any other materials.

When used in exterior walls, glass block must meet the same requirements as other materials. One factor that affects exterior walls and not interior ones is wind resistance. To be approved for outside use, a glass block panel has to resist 20 pounds of pressure per square foot, the equivalent of hurricane winds. The maximum area for glass block panels allowed by most building codes is 144 square feet—12 feet squared or a rectangle of the same or less surface area. The limit for overall height is 20 feet, while the greatest length allowed is 25 feet.

Interior walls made with glass block have far fewer restrictions. They do not need to hold back the elements, just lateral loads of 5 pounds per square foot. That means interior walls can use up to 250 square feet of glass block, or nearly twice as much as exterior walls.

In either location, glass block can only support itself. An opening for a glass block panel should have support above, just the same as for doorways or large windows. A header beam distributes the weight to the sides of the glass block, and from there to the foundation. A glass block panel that is used to support additional weight is likely to crack and fail over time.

5.5   Designing with Glass Block

Available as individual bricks, in panels, or as part of larger assemblies such as vented basement windows, glass blocks are made in many sizes, shapes, and patterns. Most are hollow, although solid blocks are made as well. Square, flat blocks of 6, 8, or 12 inches are the most common; rectangular blocks, curved shapes, angled pieces, bullnose edgers, and other variations increase your design options.

Dozens of surface patterns are available from several manufacturers, giving you almost unlimited choices for combinations of transparency and privacy. From clear surfaces to waves, ribs, stippling, and cross hatching, glass blocks offer visual variety along with safety, security, durability, aesthetic qualities, and low maintenance.

If you plan to install glass block yourself, make sure you read the manufacturer's instructions carefully. Some of the differences between glass block and other masonry products may surprise you. For example, cleaners that contain acid may work well for concrete, brick, or stone, but they can defeat the waterproofing qualities of the mortar used with glass block. And while some general rules apply to most glass block installations, such as adding horizontal reinforcement at least every 16 inches, some makers will have specialized materials, accessories, or techniques that vary from standard practices.

Before you buy the tools to install glass block on your own, contact at least two or three professional contractors. They may be able to complete the job for only a little more money than the cost of materials alone. That leaves you free to handle other tasks, and then just enjoy the results.

CHAPTER 6

Sustainability

Materials that can be produced without depleting natural supplies

Fly Ash Concrete

6.1 Building Better Blocks

First things first: cement is not the same thing as concrete. Here's the difference: cement is a substance that reacts quickly with water to form a bonding agent that holds other ingredients in place. Various forms of cement have been in use for thousands of years. The most common type of cement in use today is known as Portland cement, a mixture of limestone, clay minerals, and gypsum, heated and pulverized. Portland cement is used in grout, mortar, plaster, stucco, and concrete.

Concrete is a mixture of cement and water with fine and coarse materials, such as gravel and sand. This kind of concrete is used in exterior and interior walls, footings and floors—just about everywhere people build. In fact, more concrete is used around the world each year than any other human-made building material. Concrete depends on natural minerals for most of its substance, plus one manufactured product—cement.

As a manufactured product, cement was first made in 1824. It is not expensive or difficult to make—but another material is proving to be still more economical. Fly ash, a by-product of combustion in a coal-burning power plant, has the same properties as cement when used in concrete. It's often less expensive, and its use helps keep waste products out of landfills. One other benefit is that fly ash is ready to use, which saves energy and effort compared to making cement. This reduces manufacturing waste, too.

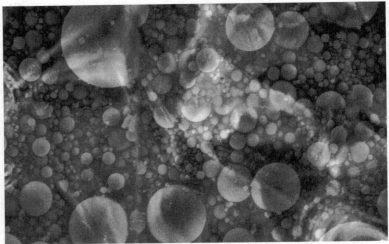

Photo courtesy of Headwaters, Inc.

6.2   Trash into Treasure

To help reduce airborne pollution, coal-burning power plants use equipment that removes ash from the plant's exhaust gases. This ash was most often added to water and diverted into "ash ponds" that cost money to create and maintain. The process of creating and disposing of fly ash was only an expense for energy producers, with no apparent benefit.

Construction industry specialists looked at the properties of fly ash and found that it could be used in place of cement in many kinds of concrete. In order to fulfill the requirements of professional builders, any substitute would have to meet standards set by ASTM International, originally known as the American Society for Testing and Materials. Fly ash does.

The money earned through the sale of fly ash for use in concrete contributes to the income of a power plant that collects and sells it. What it saves in disposal costs is even greater. What had been only a waste material has become a valuable resource in two ways. Power plants that direct fly ash toward construction use can reduce their landfill needs by 80 percent or more.

6.3   Properties of Fly Ash

Technically speaking, fly ash acts as a pozzolan—a siliceous, or siliceous and aluminous material, which in itself possesses little or no cementitious value but will, in finely divided form and in the presence of moisture, chemically react with calcium hydroxide at ordinary temperatures to form compounds possessing cementitious properties. That's how ASTM International defines it. The word "pozzolan" comes from Pozzuoli, Italy, where volcanic ash was first used in Roman times as cement in concrete.

Most environment-friendly building materials are fairly new, and their performance over time is still unknown. If you compare fly ash to volcanic ash used in concrete, though, some examples built more than 2000 years ago are still standing: structures in Cosa, Italy, that have withstood millennia of direct exposure to seawater; and the Pantheon in Rome, Italy, a pozzolan-and-lime concrete structure with a cast concrete dome 124 feet in diameter.

6.4 Benefits of Using Fly Ash

The noncombustible part of coal is left over in the form of ash with unique characteristics. In addition to its ability to react with water like Portland cement, fly ash particles are smaller than grains of cement, and are spherical, like microscopic ball bearings. These two physical features make fly ash better than cement in several ways:

- Fly ash has a lower unit weight than cement, so it contributes roughly 30 percent more volume of cementitious material per pound used; this helps the concrete flow better while it's wet.

- The smaller particles of fly ash fill in gaps within the wet concrete mixture better than cement, making the final product more consistent and therefore stronger.

- This same characteristic means concrete can be made with slightly less water and sand, coating the larger aggregates more evenly and creating a more workable material.

- While concrete made with fly ash does not equal the strength of cement-based concrete for about a month, it continues to gain strength after that, becoming more durable long after cement-based concrete has reached its maximum strength.

- Concrete made with fly ash contains less lime, and converts some of the lime present into calcium silicate hydrate; this replaces one of the weakest ingredients in concrete with one of the strongest.

- Lightweight concrete made with fly ash is easier to pump, and its smooth consistency reduces unwanted air pockets; the finished surface is more even where it is used with forms.

- Fly ash makes concrete less permeable, which improves protection against corrosion; it reduces the intrusion of water, oxygen, and chemicals, protecting steel reinforcement from corrosion and expansion.

- When cement reacts with water, it generates heat very quickly, which helps the concrete harden and gain strength; in some situations, rapid heat gain may lead to thermal cracking and loss of strength. Fly ash produces far less heat when it reacts with water, helping reduce the damaging effects of thermal cracking.

Considering all the practical benefits of using fly ash concrete, the environmental advantages may not seem as important—but they are substantial. According to industry estimates, for each ton of fly ash used instead of cement, we save as much landfill space as an average American uses in 15 months, reduce CO_2 emissions by as much as an average car produces in 2 months, and save enough electricity to power an average household for 24 days.

For the past decade in the United States, the use of fly ash in concrete has increased by more than 50 percent. At current rates, more than 12 million tons of coal fly ash are used in concrete products each year. Multiply that by the figures above, and you can see why.

One more thing: concrete itself is recyclable. Anywhere from 45 to 80 percent of crushed concrete can be used as aggregate in new construction.

6.5 Who Uses Fly Ash Concrete

Concrete made with fly ash qualifies for LEED credits from the U.S. Green Building Council. The U.S. Environmental Protection Agency requires that fly ash must be allowed on federally funded projects, and promotes the use of fly ash through its Coal Combustion Products Partnership (C2P2) program. Its use is endorsed by the U.S. Department of Energy; the U.S. Army Corps of Engineers requires fly ash concrete in most projects; and the U.S. Bureau of Reclamation has used fly ash extensively on dam projects. And all 50 states either allow or require the use of fly ash concrete in state-funded projects.

Resources

For more information about fly ash and its use in concrete, visit *www. flyash.com.*

CHAPTER 7

Renewability

Natural items that can be grown and harvested repeatedly

Bamboo and Cork Flooring

7.1 Sustaining Growth

One is hard, the other soft. Both are renewable natural materials. They look like wood, but one is a kind of grass and the other is tree bark. Their qualities make them suitable for use in floors, and they are easier than ever to find, buy, and install. These materials are different enough from traditional hardwood flooring, however, you should know more about them before deciding to use either one.

7.2 About Cork

Cork is the bark of the evergreen oak tree *Quercus suber*, which is native to southwest Europe and northwest Africa, which means almost all cork used in the United States must be imported. Most imported cork comes from the Iberian Peninsula, which includes Spain, Portugal, Andorra, and Gibraltar. Cork is also produced in Algeria, Morocco, Tunisia, Italy, and France.

While it can be grown in warmer parts of this country, it has only been successfully planted in California, where its need for direct sunlight and resistance to drought are not problems. Cork oak trees grow to be 40 to 65 feet tall, with thick layers of bark. Like other oak trees, it produces acorns as seeds. And unlike many trees that die when their bark is removed, the cork oak just grows more.

The bark can first be harvested when the tree is about 25 years old, and after that every 10 years or so. Cork oak trees live for 150 to

Photo courtesy of Sustainable Flooring, Inc.

250 years. Because each tree has a different shape, cork harvesting is traditionally done by hand.

Once harvested, cork is durable yet light and nonflammable. The cells of the bark are filled with air; they can withstand high pressures without rupturing, and will return to their previous shape. Cork has been used for insulation, packing material, and veneer, among other things. The cork we use in bulletin boards and wine bottle stoppers is the same cork used in flooring.

7.3 Cork Floor Coverings

Although only 15 percent of the cork harvested is used for wine bottle stoppers, that industry accounts for more than half of the money earned by cork producers. Understandably, wine makers get first pick of the available cork supply. Using stock that was not chosen by wine makers, as well as leftovers, flooring manufacturers grind cork into a uniform mixture. Then they add just enough adhesive to hold the material in shape, and press it into sheets. The squares and panels you see did not come straight from the tree.

Natural cork varies in color, and manufacturers sort it to produce a variety of tones. They may also add dyes to give the cork flooring a more consistent or dramatic look. Common colors range from alabaster to espresso, with most variations on a continuum between khaki and chocolate. Visual texture varies, too. Some cork products have no more directional grain than sand on a beach; others have the appearance of burl oak.

As a result of its composition, cork flooring is not strong by itself. Only when it is installed on top of a stable subfloor does it take on the strength that complements its other qualities. Remember that cork flooring is loose material formed under pressure; the larger a square or sheet of cork is, the more expensive it is to create, the more fragile it is, and the harder it is to install. That's why most cork flooring is sold in single tiles rather than rolls.

Once installed, a cork floor is comfortable underfoot, hard enough to withstand heavy foot traffic, soft enough to cushion falling objects, pleasant to look at, and resistant to fire, mold, and mildew. Its natural cushioning also adds insulation and helps control noise. With regular care, cork flooring will last for decades.

7.4   About Bamboo

Strictly speaking, bamboo is a grass—not a tree. More than 1000 species of bamboo are grown around the world. Some thrive in mountains with cold winters, while others only appear in tropical climates. For the most part, bamboo does not grow in Europe or Canada, and is seldom cultivated in other northern countries.

Where it does live naturally, bamboo grows very quickly. Some of the largest species grow fastest of all, with new shoots gaining 3 feet per day. Bamboo is one of the only plants whose rapid growth can be heard and seen as it happens.

The genus *Phyllostachys* includes several large species of bamboo that are large enough to use in building products. These are known as *timber bamboo*. In addition to growing vertically, these types of bamboo spread quickly through rhizomes under the surface of the soil. In fact, they are even considered invasive species in places like California, where they are not native. Of the roughly 75 species and 200 varieties of *Phyllostachys*, the largest can grow to be about 100 feet tall.

To minimize impact on wild habitats and maintain consistency in the harvested product, some companies that make bamboo products have established their own plantations. Whether cultivated or collected, mature bamboo can be harvested every five years or so.

7.5   Bamboo Floor Coverings

By itself, bamboo is not a suitable replacement for wood in flooring. Individual stalks of bamboo can be used in building, but flooring has to be broad and stable. To become a useful flooring material, bamboo must be shredded and then formed into regular shapes, such as planks and tongue-and-groove boards, under high pressure. As with cork, this loose material is held together with an adhesive or resin.

The adhesives and resins used to create bamboo flooring, transform an abundant natural resource into a durable building product. In keeping with the goal of environmental responsibility, manufacturers emphasize the use of adhesives that do not contain formaldehyde, and resins such as acrylic urethane that contain few or no volatile organic compounds. This helps the products meet standards for both green manufacturing practices and indoor air quality.

Once shredded bamboo is set in its new shape, it provides a flooring surface that is harder than red oak or maple, and thick enough to refinish several times. In tongue-and-groove boards, bamboo is available with vertical or horizontal grain, naturally colored or stained, finished or unfinished. Bamboo flooring is also made as a laminate, with a $1/8$-inch wear layer of bamboo adhered to a backing made of pine or other softwood.

7.6   Buying Tips

The increasing popularity of cork and bamboo has brought many manufacturers to the field, and their competing claims can be confusing. When you shop for renewable flooring materials, check more than just the price per square foot. Think about whether the material will still be appropriate if you change the way you use that room. Finally, ask installers which products they believe stand up best over time. Your investment in renewable materials will benefit the environment; make sure it also improves your quality of life.

CHAPTER 8

Recyclability

Materials that can be broken down and used again in another form

Plastic Plumbing Pipe

8.1 Creating New Standards

As we know them, plastics are polymers that usually include petroleum. Because many green initiatives encourage us to reduce our use of petroleum products, plastic plumbing pipe may seem out of place in a list of environment-friendly materials. Compared to the total costs of using copper, however, materials such as polypropylene and cross-linked polyethylene (PEX) become much more attractive, with benefits that traditional materials do not offer.

Two polymers currently in use for water supply lines are PEX and polypropylene. PEX resists extreme temperatures and handles most liquids without reaction, which makes it a smart choice for indoor plumbing. Unlike traditional copper pipe, PEX can bend to follow curves and work around corners. Polypropylene provides many of the same benefits as PEX, and is recyclable as well.

One polymer used in plumbing pipe, polyvinyl chloride (PVC), is associated with environmental harm. Water mains made with PVC before 1977 contain high levels of vinyl chloride monomer, a toxic chemical known to be a carcinogen. Newer PVC pipes contain lower levels of harmful compounds, but the manufacturing of PVC still produces dioxins, and the use of PVC in plumbing is generally restricted to drain lines.

Both polyethylene and polypropylene are made without chlorine, bromine, or other harmful chemicals. They are considered benign plastics.

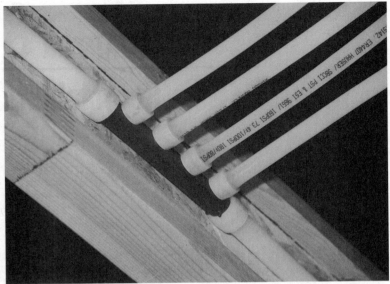

Photo courtesy of Plastic Pipe and Fittings Association

8.2   About PEX

PEX is a form of polyethylene in which the molecules are cross-linked to increase strength and durability without sacrificing flexibility. This makes it easy to install, but difficult or impossible to recycle. Still, PEX offers sufficient environmental benefits to be considered a green alternative to copper pipe. Even as polypropylene pipe becomes more common in the United States, PEX will remain a viable option, thanks to its ease of use and low cost.

Since the 1960s, PEX has been used in plumbing and radiant heating systems in Europe. It was first introduced in the United States in the 1980s, and its use has grown steadily since then. It's flexible in more than one way; because PEX handles water at all temperatures from just above boiling to just below freezing, it is suitable for use in home plumbing, service lines, hydronic heating systems, and snow-melting applications.

In home plumbing, PEX can be routed much more easily than copper pipe. This makes remodeling projects far simpler, because it can be worked around other systems such as drain lines, gas pipes, electrical conduit, and so on. Because PEX can curve and bend, it requires fewer joints, which reduces the potential for leaks.

The inside diameter of PEX pipe is relatively small, but it has a low coefficient of drag. This means liquids flow smoothly through it without leaving mineral deposits, and less standing water is left in the pipes, so you can adjust faucet temperatures more quickly. PEX is also quieter than copper pipes.

8.3   Finding and Buying PEX

Thanks to its growing popularity, PEX is now available through many retail building supply stores. If the retailers closest to you don't carry PEX, look for a plumbing supply wholesaler, who may be willing to sell directly to you—even if you don't have a contractor's license. The same seller who carries PEX pipe should also be able to provide the fittings you need.

Sizes from ¼ inch to 1 inch are common; larger diameters are available, but most of the time only as special-order items. The sizes of PEX pipe are given in copper tube size (CTS) numbers, based on the sizes of comparable copper tubing. All sizes should have the same pressure ratings, because the wall thickness is proportionate to the diameter.

Shorter lengths are sold as straight pipes up to 20 feet long, while coils of up to 1000 feet provide enough length for many separate lines. Some manufacturers color hot and cold lines differently; make sure the pipe you buy is rated for its intended use.

8.4   Using PEX

From lower material costs to easier installations to reduced maintenance, PEX can help you save money, time, and effort. If you install a manifold at the source, with cold lines running directly from the municipal water supply and hot lines from the water heater, you can set up separate water lines for each room—or even each fixture. This "home run" configuration brings hot water to the fixtures more quickly, saving both water and the energy used to heat it.

Speaking of heat, you can use PEX for hydronic systems. It can be encased in concrete for use as part of an in-floor heating system. Its flexibility allows you to create a single, continuous loop of pipe with no joints or connectors. At the other end of the thermometer, you can run very cold water through PEX pipe; just don't use it for outdoor lines above ground. Exposure to direct sunlight and severe weather can shorten the life of PEX pipe.

To connect separate pieces of PEX, installers use mechanical insert fittings, either crimp style or compression fittings. In part because of its resistance to friction, PEX should not be fastened with glue or heat. For a list of approved connectors, check with the manufacturer or review the standards ASTM F 1807 and ASTM F 1960. One advantage of mechanical fittings is that you can test them immediately, without waiting for solder to cool or glue to dry.

Before you specify or install PEX as plumbing pipe, check your local building codes. If you need to provide evidence that PEX is safe for use in home plumbing, you'll have plenty of support. It conforms to the standards ASTM F 876 and F 877, AWWA C904, and CSA B137.5. Polyethylene pipe has been tested and approved by ASTM, ANSI/NSF

International, and CSA, and all major model plumbing codes used in the United States and Canada; NPC, UPC, IPC, and NSPC.

8.5 About Polypropylene

While water from polyethylene pipe is safe to drink, some people say it imparts a taste they notice and dislike. Polypropylene pipe, on the other hand, does not contain the chemical stabilizers that are common in polyethylene. Its chief advantage over polyethylene, though, is not how it's made but that it can be recycled.

The same properties that make PEX pipe a worthy choice are also true of polypropylene. It's flexible, economical, and easy to install, and it does not react to common household chemicals. Like polyethylene, polypropylene has a low coefficient of friction so fluid moves easily through it, and it's slow to conduct heat, which means the water temperature won't change much in transit. Watch for polypropylene pipe to become more common in the next few years.

CHAPTER 9

Simplicity

Products that require little processing before they are used

Red Cedar Siding

9.1 Following Tradition

Not many materials go from raw natural resources to finished products with fewer steps than cedar. Even the mud that covers adobe walls requires more handling. With cedar, a tree is felled, barked, cut, and dried; that's it. And cedar has uses inside and outside the house, all of which provide livability upgrades while remaining environmentally responsible.

For one thing, few other materials can be created again. The ingredients used to make metals, concrete, plastics, and other building products depend on limited natural resources. All it takes to make more wood is land, seeds, and time. Unlike other extractive industries, timber growers are able to reinvest in future production. Forest-products companies large and small understand that their stewardship today will determine their success tomorrow. According to industry groups, North American forests have grown 20 percent since 1970.

9.2 Benefits of Using Cedar

As you may remember from Chap. 1, nearly half of the industrial materials manufactured in the United States are wood products. To produce that much material, you might expect the wood products industry to consume nearly half the energy used in manufacturing. But in case you missed that detail, here it is again: production of wood products takes up roughly four percent of the energy used in making

iStockPhoto

building materials. Making steel consumes more than twice the energy, and making concrete consumes at least four times as much.

The LEED system, mentioned in Chap. 4, uses several criteria for judging a building project. The principles underneath those criteria are reducing energy use, reducing resource use, minimizing pollution, and reducing environmental impact. Wood products perform exceptionally well in all four areas.

The Athena Sustainable Materials Incentive in Canada compared the environmental effects of wood with those of steel and concrete. Wood had the lowest impact on total energy use, greenhouse gases, air pollution, water pollution, solid waste, and what they termed ecological resource use. Wood is also less expensive to transport to or use at a building site.

One more unique property of wood products: during the entire time it takes for a tree to mature, it consumes carbon dioxide and produces oxygen. The environmental benefits of using cedar begin the moment each new tree is planted. The benefits continue from there.

9.3   Cedar as a Construction Material

Western Red Cedar is dimensionally stable; it seldom changes shape through shrinking, swelling, or warping in all climates. This stability is unique among softwoods. It holds nails, screws, and glue well, and is easy to work with. Without other treatment, cedar resists moisture, insects, and decay.

Roughly half the energy consumed in the average house goes to space conditioning—heating, cooling, and ventilating. This makes insulation a high priority for controlling energy expenses, especially in climates with extreme temperatures. Wood is an excellent insulator, making it ideal for framing. However, despite its natural resistance to decay and insects, cedar is not as well suited as other species, such as Douglas fir or southern yellow pine, for building.

Priced by the board foot, cedar is more expensive than pressure-treated lumber, in part because it is more pleasant on the eyes and the nose. Concealing it inside walls would only hide some of its most desirable features. Its aromatic properties and hearty appearance make it ideal for lining closets, encasing exposed beams, framing windows, and so on. Even these are uncommon, though. Most of the cedar used for building stays outside.

You can use cedar for a variety of outdoor building projects, from decorative items to the very surface of your house. Other outdoor projects that make good use of cedar include fences, gates, arbors, trellises, and gazebos.

One thing that applies to all outdoor cedar projects held together by nails or screws: use only corrosion-resistant fasteners, such as aluminum, hot-dipped galvanized, or stainless steel. Stainless steel, while more expensive, is generally considered the best choice. All other metals are likely to corrode, leaving stains that diminish the look of the cedar.

9.4 Cedar Siding

The most popular styles of cedar siding are bevel, pattern, and board-and-batten. The style you choose will depend on your preferences and the design of your house.

Bevel siding, the cedar siding most often used, is tapered from one edge to the other. One face has a rough-sawn texture; the other can be rough or smooth. This style of siding is always installed horizontally. The cost of one variety over another depends on the clarity of the wood—that is, how many visible knots and other growth characteristics can be seen—and the length of each piece. Continuous lengths cost more than finger-jointed boards.

Pattern siding includes both tongue-and-groove and lap styles. Tongue-and-groove boards are milled on both edges; the top edge has a thin, narrow shape, which matches a channel cut out of the bottom edge. A few variations in shape are available, but this is one place style doesn't matter; once assembled, both the tongue and the groove are hidden from view. These shapes interlock to create a continuous cedar surface, which means they can be installed in any direction the framing will support. The grade of wood used will depend on the desired effect, from formal to rustic.

Lap siding boards are similar in the way they fit together; the main difference is that each new board rests on top of the one before, rather than fitting into it. Like bevel siding, lap boards are rough on one side and rough or smooth on the other. And like tongue-and-groove siding, lap boards can be installed in any direction. The most common sizes range from 1 × 6 inches to 1 × 10.

The board-and-batten style uses a surface of wide boards fitted side by side, with narrower boards, called battens, covering the joints. While this pattern could be installed horizontally, it is most often vertical. The widths of both the boards and the battens will depend on the desired look, which in turn depends on the size and style of the house. One variation with a dramatic effect is installing the battens first, creating deep channels between the boards.

9.5   Finishing Cedar

Left unfinished, cedar ages to a distinguished silver gray. This look is well suited for fences, gates, trellises, gazebos, and other free-standing features. But applying a surface finish will help those and all other cedar projects look better and last longer.

Contrary to older notions about cedar, it should not be left to weather. Once it does, the cedar will not absorb protective finishes as easily or hold them as well. This can lead to premature peeling, blistering, or cracking in the finish. The longer you wait to finish new cedar, the less effective the finish will be.

To preserve the freshly installed look of cedar, use a finish that includes protection against ultraviolet light—the chief cause of discoloration. Cedar used in wet climates will also benefit from mildew protection, which may be available in the same finishes.

If you want the weathered look but prefer to keep the wood protected, apply a bleaching stain that is formulated for use with cedar. It uses chemicals to create an appearance of weathered aging, but preserves the integrity of the wood at the same time. Like other finishes, bleaching stain should be reapplied at regular intervals. Remember that the point is to protect the wood—and your investment.

CHAPTER 10

Efficiency

Items that consume less energy in production and use

Recycled Insulation

10.1 Reducing Waste

Whether you use electricity, natural gas, oil, or another energy source, at least half of the fuel consumed goes for space conditioning—in other words, heating and cooling. That's because some of the indoor temperature leaks outside, and the outside temperature leaks in. To gain better control over your indoor atmosphere and reduce your energy costs, make sure you have the right kind of insulation, and enough of it in the right places.

10.2 Choosing an Insulating Material

When most homeowners think about insulation, they picture fiberglass batts. This makes sense, because fiberglass insulation is practical, affordable, and relatively easy to handle. By comparison, loose fill insulation involves more effort and sometimes more expense. Once the insulation is installed, however, loose fill has several advantages that make it worth the trouble.

The first of these is recycled content. Loose fill insulation can be made from

- *Recycled paper:* The most abundant material available for recycling.
- *Mineral wool:* Either the manufactured material called rock wool or the metal by-product slag wool.
- *Fiberglass:* It can be reclaimed from previously discarded insulation.

Photo courtesy of GreenFiber

Loose fill is either blown in to confined areas such as attics and crawl spaces, damp-sprayed into wall cavities, or poured into areas where access is easier. Moisture is sometimes added to the loose fill to keep the material from blowing where it doesn't belong. This evaporates quickly once the insulation has settled.

One weakness of cellulose insulation is that it can absorb more moisture than most other types of insulation. If it doesn't have a chance to dry out again, it will lose some of its effectiveness and perhaps become a place for mildew to grow.

10.3   Reducing Waste

In some ways, loose fill is only on par with fiberglass batts. For example, they all have roughly the same fire ratings, and cost about the same for the amount of R-value they provide. Factor in the recycled content, and the balance tips toward loose fill. Then take into account the potential health hazards, and the recycled products look better still. For the greatest number of environmental benefits, consider cellulose insulation.

Not only is cellulose insulation made from up to 80 percent recycled fiber, it can be recycled again. By using cellulose loose fill, you keep unwanted air and sound out of spaces in your house, and keep cardboard and paperboard out of landfills.

Made from paper products and wood fiber, cellulose insulation fills in spaces large and small. It can be packed into tight places, such as the points at which the roof decking and rafter ends meet the ceiling joists. It can be laid down in thick blankets where there's more room, and gravity is on your side. The insulating value of cellulose loose fill is about R-3.7 per inch. It guards well against sound transfer, too.

10.4 Protecting Health

The binders used in most fiberglass insulation contain formaldehyde, which is widely considered harmful to your health; its vapors are classified as potential carcinogens. Once installed, fiberglass can still emit formaldehyde gases. Fiberglass is also listed as unhealthy, both for the long-term effects of exposure to airborne particles, which some sources consider another possible cause of cancer, and the immediate result of respiratory irritation.

To make cellulose loose fill more resistant to fire and infestation, manufacturers use chemical additives. Up to 20 percent of cellulose insulation is made of ammonium sulfate or boric acid, both of which are low in toxicity. Ammonium sulfate gives off a stronger odor, and some installers believe it may corrode copper pipes. Boric acid, meanwhile, is also used in baby clothes and cotton batts as a fire retardant; it is less likely to cause irritation.

If you're concerned about the potential health hazards of these additives, either during installation or afterward, ask the manufacturer or retailer for health and safety information—or check for competing products.

10.5 Hidden Hazards

When you evaluate your existing insulation, watch out for vermiculite— a lightweight mineral shaped like grains or nuggets, in colors that range from gray-brown to silver-gold.

Vermiculite was commonly used as loose fill insulation in houses built before 1990. Some contain asbestos, which is well known as a carcinogen. If asbestos fibers are released into the air and inhaled, they can lead to serious respiratory illness or cancer. The threat to your health is great enough that the U.S. Environmental Protection Agency (EPA) has issued a warning about vermiculite insulation.

If you find vermiculite in your house, call an insulation professional for advice. You may have to decide between having it removed by a licensed contractor and just leaving it where it is. For more information, visit the EPA web site at *www.epa.gov and type "vermiculite" in the search box.*

10.6 Installing Loose Fill

While fiberglass batts have the same R-value as cellulose loose fill, they are made in fixed widths and thicknesses. Any variation in the space you fill with fiberglass batts may prevent them from forming an adequate barrier to air infiltration and high-frequency sound transmission. Remember: the R-value of a gap in your insulation is zero.

Loose fill, however, conforms to every bump and dent, sealing the space with as much material as you choose to apply. Over the months and years, cellulose insulation settles further in, which improves its thermal performance. That also means the installed depth must be adjusted for the predicted density once the loose fill has settled, not just the total depth at the time of installation. Don't worry about making these calculations; manufacturers provide them in their product literature, and sometimes right on the bag.

For installations in open wall cavities, you will need to use a damp-spray method. While this may be within the range of skill you already have, the requirements are strict and the consequences of mistakes can be serious. The dampened mixture should not have more than 25 percent moisture when it is enclosed, and will need enough ventilation to finish drying out. Otherwise, mold and mildew become likely. This is one project you should probably give to a professional installer.

The installed price of cellulose insulation can be higher than for fiberglass batts, because loose fill requires special equipment to reach tight spaces. As a percentage of total building or remodeling costs, however, this difference in price is relatively small. Given the performance characteristics of loose fill, you will save that much and more in heating costs over time, without even calculating the value of environmental and health benefits.

10.7   Finding Your Number

To determine how much insulation you need, ask your local housing inspection agency or power utility for recommendations. Another valuable source of information is the Home Energy Saver, an online calculator created by the Lawrence Berkeley National Laboratory. It's available at *http://hes.lbl.gov.*

PART II

Technology Changes

New Ways of Meeting Your Household Needs

When you replace or remodel parts of your house, your first impulse may be to get another or more of what you already had. Unlike new construction, repair and renovation have to fit with what's still there; these situations don't often inspire people to experiment with new concepts. On the other hand, a home improvement project can be just the right scale for trying something unfamiliar. Having radiant in-floor heat in just one room, for example, can give you an idea how well you like it, and whether it works for the way you live.

Features and Functions of Daily Life

In this part of the book, you will learn about different ways of meeting the basic needs of shelter—comfort, safety, and protection from the elements, and a place that feels right to you. Some of the technologies described here bring entirely new ideas home; others just use technology to simplify the way we use and conserve energy and natural resources.

11. Electrical 1: Power Consumption Monitors
12. Electrical 2: Timers and Motion Sensors
13. Plumbing 1: Low-Flow Plumbing Fixtures
14. Plumbing 2: On-Demand Water Heaters
15. Heating: Radiant In-Floor Heating
16. Ventilating: Heat and Energy Recovery Ventilators

Another Step Forward

Part 1 described several options that take the place of familiar materials. With the exceptions of structural insulated panels and polyethylene plumbing pipe, these materials even look and feel a lot like the products they are meant to replace. The chief difference is in their environmental performance, according to the criteria listed at the beginning.

In Part 2, you will learn about recent technological developments that go a little further. Some of the products in the following 10 chapters are the latest versions of established devices, such as induction cooktops or photo-controlled light switches; others represent large changes, or even completely new approaches, in residential construction.

By the time you finish this part, you should have at least one or two ideas to pursue. Move ahead with confidence; these technologies are ready for use.

CHAPTER **11**

Electrical 1

Power Consumption Monitors

11.1   Keeping Tabs

People who heat their houses with wood, coal, heating oil, or LP gas generally know how much fuel they consume. They see how often they need to restock, and are reminded of the price each time. Natural gas and electricity, however, are delivered constantly and on demand. That's convenient for homeowners, but they can lose sight of how much they use.

In order to save on utility costs and reduce your consumption of natural resources, you need a better picture of your energy use habits. Because you directly control most of the appliances and fixtures that use natural gas, you can estimate what percentages are consumed by your range, boiler or furnace, clothes dryer, water heater, and so on. But electrical consumption is harder to guess, and impossible to see.

Until recently, you could review your total use of electricity and estimate how much went to each appliance or fixture. Larger items such as refrigerators and washing machines have carried Energy Star ratings for several years, so their consumption is easier to calculate. Items like baseboard heaters are also simple to identify as heavy users. Knowing whether your kitchen, family room, or home office used more electricity was pure guesswork.

Thanks to the introduction of recent products, you can now measure your electricity use more accurately. Some show real-time consumption of your entire system to give you immediate feedback on what you're doing; others measure the rate of consumption so you know which fixtures use the most current. These devices show where the electricity goes, the way a check register shows where you spend

Photo courtesy of P3 International

your money. Like check registers, these products give you a starting point for creating a budget and taking control of your household.

11.2   Getting the Big Picture

A product called the PowerCost Monitor shows how much electricity your house is consuming in real time. It has two parts: a sensor and a display.

The sensor, which is designed to work with all standard digital and electromechanical meters in North America, is fastened to an existing household utility meter with a ring clamp. Adding the sensor will not change the way the meter works; it is only attached to the outside. The display, which you can place anywhere you like, receives signals wirelessly from the transmitter and shows consumption information in kilowatt hours—and in dollars and cents. The display also shows other information, such as the time and outside temperature.

According to the manufacturer, real-time feedback helps homeowners reduce their consumption of electricity by 10 to 20 percent. As with any measuring device, the PowerCost Monitor only shows the results of your actions; you still have to decide what things you want to do differently. Information from a monitoring system like this can help you determine the effect that different actions will have on power use, and on your electricity bill.

Another whole-house power monitor is called The Energy Detective, or TED. This device reads overall energy consumption from within the house, rather than on a meter mounted to the exterior.

It includes a chart that describes its functions and provides tips on saving energy. The chart shows current electrical use in watt seconds, daily peaks, and totals, as well as monthly usage, in kilowatts and in dollars.

Like the PowerCost Monitor, TED has two parts—a transmitting sensor and a display. The sensor is attached to your electrical service panel, and the receiver with display can be plugged into any outlet in the house. Because installation requires direct contact with the service panel, only qualified homeowners and electricians should install it.

TED also calculates nighttime loads—how much energy is being consumed when switchable appliances are turned off. This shows the consumption rate of appliances and fixtures that remain on all the time.

One more device for measuring electrical consumption, this one from Australia, is the Cent-A-Meter. In addition to electrical current, it displays in-door ambient temperature, humidity, and the equivalent greenhouse gases generated in producing the power consumed. Its functions are similar to those of the PowerCost Monitor and The Energy Detective.

11.3   Seeing the Details

A product called the Kill-A-Watt EZ Plug Power Meter measures electricity use by plug-in appliances, showing the rate of consumption in watt hours and kilowatt hours. It goes between an electrical outlet and the appliance it powers, displaying how much electricity each item consumes. In addition, it measures the quality of your power source by checking the line frequency, voltage, and power factor.

The device also shows running and projected costs. It lets you see what each appliance costs to run for a given period—day, week, month, or year. This lets you distinguish between constant costs for items that run all the time, such as refrigerators, and those you only run some of the time, such as air conditioners.

A similar device called the Watts Up Pro Portable Plug In Power Meter offers roughly the same functions, with optional software that lets you create charts of your electrical use. Like the Kill-A-Watt EZ, the Watts Up Pro shows consumption in watt hours, with automatic conversion for cost, based on a rate you enter. The Watts Up Pro Monitor shows

- Current watts
- Minimum watts
- Maximum watts
- Power factor
- Cumulative watt hours

- Average monthly kilowatt hours
- Tier 2 kilowatt hour threshold (used to calculate secondary kilowatt hour rates)
- Elapsed time
- Cumulative cost
- Average monthly cost
- Line volts
- Minimum volts
- Maximum volts
- Current amperes
- Minimum amperes
- Maximum amperes
- Power cycle

This seems like more information than most homeowners need, but customers who use these devices become curious about the details of their energy use, and apply that knowledge to better manage their households.

The Watts Up Pro stores data in nonvolatile memory, which you can retrieve even after a power outage. It samples use over time, for up to a thousand data points, so you can download the information to your computer and see trends in your power consumption. You can also export the data in comma-delimited format for use in any spreadsheet program.

11.4   Predicting the Future

Because the Kill-A-Watt EZ and Watts Up Pro are portable, you can take them along when you shop for new appliances. If a demonstration model of an appliance you consider buying is plugged in, put the power monitor in the circuit and see how it compares to your existing appliance. The Watts Up Pro even has a payback calculator that figures the time required for an energy-efficient appliance to pay for itself. It shows monthly savings compared to the purchase price of the new appliance.

All these devices measure the way you use electricity; none will control any appliances or make any decisions for you. Once you have a clear picture of where your energy money goes, though, you will have a much better idea what to do.

CHAPTER 12

Electrical 2

Automating your use of electricity

Timers and Motion Sensors

12.1 Improving Safety and Security

No matter how you measure your consumption of electricity, you know how to save on your power bill: turn appliances off when you're not using them. Saying it is easy, but doing it is sometimes a challenge. For example, you may have family members who don't pay attention to the use of electricity, or you may be so busy yourself that even one more thing to think about is one too many.

Automatic switches, whether controlled by clocks, photo cells, or motion sensors, won't do your thinking for you—but they will turn electric items on and off without your actions. They can help minimize the unnecessary use of electricity in your house while improving your comfort, safety, and security. The amounts you save using these devices will quickly pay for them; after that, it's money you get to keep.

12.2 Photo-Controlled Switches

In winter months, the work day is longer than the daylight. You might leave home and return in the dark without changing your schedule. But just as you wouldn't leave your thermostat set to full warmth while you're gone, you wouldn't leave the lights on all day.

A simple photo-controlled switch will turn on a light when the surroundings become dark. Early models often flickered, because the light that was switched on would affect the photo cell. This is no longer a problem. A modern photo control turns lights on and off as

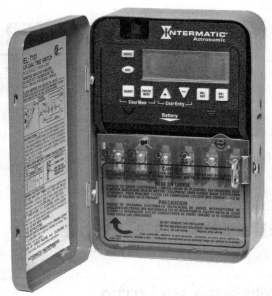

Photo courtesy of Intermatic, Inc.

surely as a hard-wired switch. At least one model currently available turns lights on at dusk, and off again six hours later, so the light won't be left on overnight.

Just one photo-controlled switch, paired with a lamp near the door you normally enter, can help welcome you home. Two photo-controlled switches can help you illuminate front and back doors, or the main entry plus an interior room, making your house look occupied. This extends the benefits from safety to security.

12.3 Indoor Timers

For appliances other than lights, your needs are less likely to depend on the changing daylight hours. For example, you may need to run a dehumidifier while you're out, or start an air conditioner an hour before you expect to get home. For tasks like these, a timer is a better choice.

Timers are available in analog and digital models, with anywhere from one on-off cycle per 24 hours to as many as a cycle every hour. Different models can handle varying current loads; the most basic timers will control light fixtures, while heavy-duty grounded models handle high-demand appliances such as air conditioners and heaters. Special models are made for 3-prong, 220- to 240-volt grounded appliances.

Most timers plug directly into the wall, but some have cords and table-top controllers. These models are designed to help users who may not be able to reach an outlet. Others can be mounted in the wall, in place of switch boxes, for easy access with no cords to manage.

As timers have become more reliable and versatile, they have also become more affordable. Even the most basic models allow manual overrides without disabling the timer. Many offer random pattern switching for greater security. This feature varies the time at which a light or other appliance is switched on and off, so anyone watching your house won't notice the same things happening at the same times every day.

For an automated switch that uses a timer and tracks the cycles of daylight, you can now find models with "astronomic" features. That doesn't refer to the price, but to the timer's capability to follow a year-long calendar indicating sunrise and sunset times for each day.

12.4   Outdoor Timers

If you have outdoor lighting, an outdoor timer can help you save electricity by leaving the current off until it's wanted. For example, if you use 110 to 120 volt current to power walkway lights for greeting visitors, you won't want the lights to stay on all day. Or if you have a water feature to attract wildlife, but only want the pump running during certain hours, a timer will save you the trouble of remembering to switch it on and off. Or if you enjoy putting up holiday lights, you want to keep your tradition going while saving energy.

Fewer outdoor timers are available than indoor models, but any building supply store will have several from which to choose.

12.5   Power Failure Lights

In case of a power failure at night, you will need a way to see. Special lights designed for just this purpose stay plugged in, drawing a small amount of current. When normal power is interrupted, the lights switch on automatically, giving you a beacon in the dark. The most likely place for a power failure light is near the master bedroom; you might also consider placing one in a stairway, hallway, kitchen, or bathroom.

12.6   Motion Sensor Switches

Already common in neighborhoods across the country, motion sensor switches switch on outdoor lights only when a large object moves nearby. These have become popular for both energy savings and security; in fact, they are often given away by crime prevention programs. A basic model replaces an outdoor light fixture, and has a sensor and one or two bulb sockets that mount directly to the electrical box. Most motion sensor switches are designed for outdoor use; they should have rubberized gaskets to protect the electrical box and the bulb sockets.

Despite their value in improving safety and security, most outdoor lights with motion sensor switches have one flaw: The lights usually point directly at the people moving toward the house. To prevent this inconvenience, aim the bulbs outward from the fixture, or use a separate motion sensor and mount the lights so that they illuminate the entryway, not the approach.

12.7   Home Automation Controllers and Modules

If you like the idea of automating your use of electricity, and have more than two or three devices you would like to control with timers or sensors, consider investing in a home automation system. Until a few years ago, these were complicated and expensive, requiring professional installation and specialized equipment. You can now find do-it-yourself home automation kits with comprehensive programming options for surprisingly little money.

Using a single controller, you can create and change settings for as many items as you like—up to 16 devices on each of 12 channels. The system has modules for outlets, sockets, and wall switches, all receiving commands from the controller through radio frequencies. That means you don't have to run any new wiring.

Such a system includes the most advanced features found in timers: programmability, manual overrides, multiple on-off cycles, random timing for security, automatic adjustment for daylight savings time, multiple dimming levels, and astronomical figures. It even has self-diagnostic functions so you can be sure it's set up correctly.

12.8   Now and in the Future

The payback period for energy-saving devices will depend on the cost of the resource, the price of the device, and how much it helps you save. The cost of electricity is increasing, and the prices of motion sensors, photo controls, timers, and home automation systems have come down within reach of most homeowners. If you need to leave any electric lights or appliances running when you're not home, chances are you'll earn back the price of any of these items—and soon.

CHAPTER 13

Plumbing 1

Reducing your water consumption

Low-Flow Plumbing Fixtures

13.1 Slowing the Flow

More of our blue planet is covered with water than with land—yet only about one percent of that water is available to us. Fresh water is in greater demand than ever, while the supply is not changing. Along with household uses such as drinking, bathing, and washing, water is used on lawns and gardens. All these uses add up, and waste increases the numbers.

Two converging trends point to likely water shortages in parts of the United States. Areas with higher population growth also have higher per capita water use. More people using more water each will accelerate the pace of consumption. Reservoirs in western states are already running low, and we have no way yet to replenish them. The best way to ensure an adequate supply of fresh water in the future is to conserve it now.

13.2 Having WaterSense

The standard rating system Energy Star helps consumers see two things on every labeled major appliance: how much the average operating cost will be, and how a given product compares to others in its class. Now, a similar system provides common points of reference for homeowners as they look for plumbing fixtures that save water.

The WaterSense program from the U.S. Environmental Protection Agency (EPA) establishes criteria for water efficiency and performance. Each product category has a different set of testing and certification

iStockPhoto

protocols. These serve as benchmarks for licensed third-party groups to use in testing. Products that meet all criteria for their category are allowed to use the WaterSense label. The program helps consumers find products that meet their expectations for performance and efficiency; it also encourages manufacturers to continue looking for ways to create better products.

To participate in the WaterSense program, each manufacturer must form a partnership agreement with the EPA. If the agency has established a specification for a category in which a company makes a product—for example, garbage disposal units—the manufacturer has one year to pursue certification for that product. The manufacturer must then have the product certified by an independent, licensed body approved by the EPA.

Product categories that can currently earn WaterSense certification include home plumbing fixtures such as toilets, bathroom sink faucets, and showerheads. Landscape irrigation services, including drip irrigation systems and irrigation control technologies that use sensors or weather-monitoring systems, may also qualify. And, because so many plumbing products are made for commercial and industrial use, the EPA has categories for commercial valve-type toilets, urinals, and steam sterilizers.

13.3 WaterSense Basics

WaterSense began as the result of meetings held in 2004 by the EPA. In those meetings, agency officials asked various groups for suggestions on creating a voluntary national program to promote the manufacture and use of water-efficient products. WaterSense was launched in 2005 with a specification development process for high-efficiency toilets (HETs) and with criteria for endorsing certification programs for irrigation professionals.

Since then, the EPA has used a set of principles from WaterSense as it selects products for evaluation, develops product specifications, and makes choices about label use, partner status, leader recognition, and marketing efforts. In order to qualify for the WaterSense label, products must

- Perform as well as, or better than, their less efficient counterparts
- Be about 20 percent more water-efficient than average products in the category
- Realize water savings on a national level
- Provide measurable results
- Achieve water efficiency through several technology options
- Be effectively differentiated by the WaterSense label
- Be independently certified

13.4   Certified Products

Most of the plumbing products we use are in the bathroom and the kitchen. Most households also have plumbing for a washing machine, and some use water in radiant heating systems. For the majority of homeowners, though, the bathroom has the greatest concentration of plumbing fixtures, so this is where the WaterSense program was originally focused.

The first three product categories to be evaluated for improvement criteria were toilets, bathroom sink faucets, and showerheads. The requirements for high-efficiency toilets are established, while those for bathroom sink faucets and showerheads are forthcoming.

High-Efficiency Toilets

Toilets account for roughly 30 percent of residential indoor water consumption. According to WaterSense, inefficient toilets are responsible for most of the water wasted in American homes. That inefficiency can be the result of leaking, excessive water use by design, or both. For several years, federal law has required new toilets to use no more than 1.6 gallons per flush. The law does not apply to older toilets, though, which may still use much more.

From 1980 to 1994, the law required toilets to use no more than 3.5 gallons per flush (gpf). In 1994, that was reduced to 1.6 gpf for two reasons: the need to reduce water consumption in order to sustain our supply was becoming clear, and most manufacturers had found ways to make toilets that worked correctly using less than half the water that older models consumed. Before 1980, little attention was paid to water consumption by toilets, and some models used as much as 5 gpf. Replacing one of these would bring immediate and substantial savings.

Leaking is also a serious problem because it can be difficult to detect. Like the slow drip from a faucet, a leaking toilet can consume hundreds of gallons of water every year, adding to your water bill and reducing the available supply. However, a leaking toilet is easy to fix with inexpensive replacement parts, which you can find at any hardware store or building supply center.

The best way to solve both problems is to install a newer toilet, particularly one that meets WaterSense requirements. These high-efficiency toilets go beyond the expectations of current laws. Instead of using a maximum of 1.6 gallons per flush, they are limited to 1.3 gallons. That's less than a 20 percent improvement in water consumption—but just like leaks, those small amounts add up quickly. The WaterSense program estimates that replacing older, less efficient toilets with HETs would reduce our national water consumption by 2 billion gallons per day. The agency puts it in dramatic terms: "If every home in the United States replaced one old toilet with a new HET, we would save more than 900 billion gallons of water per year, equal to more than two weeks of flow over Niagara Falls!"

The benefits for each household are significant over time, too. The WaterSense program calculates that if you install a certified HET, you can save 4000 gallons per year. Based on an estimate of 140,000 flushes in an average lifetime, each of your children can save as much as 300,000 gallons using an HET installed now.

Translating gallons into dollars, the savings are easier to understand. A family of four that replaces a 3.5 gpf toilet with a certified HET can save more than $90 a year on water bills, and $2000 over the lifetime of the toilet. This means a new WaterSense labeled HET can pay for itself in only a few years—sooner if the local water utility offers incentives for replacing older toilets with HETs.

Of course, a toilet that uses less water will only seem like an improvement if it works as well as an older model that uses more water. That has been one of the requirements of the WaterSense program. Besides, manufacturers know that if the first hundred HETs don't perform well, they won't sell the second hundred. Many HETs have proven better than standard toilets in consumer testing.

Bathroom Sink Faucets

In the United States, more than a trillion gallons of water are consumed through faucets each year. That represents more than 15 percent of indoor household water use. Current laws require faucets to consume no more than 2.2 gallons per minute (gpm), but older faucets may flow at rates anywhere from 3 to 7 gpm. High-efficiency bathroom sink faucets and accessories, such as aerators, can help reduce consumption by 30 percent or more.

Bathroom sink faucets that earn WaterSense certification are required to perform as well as the high-flow models they replace—even in houses with lower water pressures—but use no more than 1.5 gpm. By installing at least one WaterSense certified faucet or aerator in each household in the country, we could reduce our annual national water consumption by more than 60 billion gallons.

Showerheads

About 17 percent of residential indoor water use goes to showering, which adds up to more than 1.2 trillion gallons of water consumed each year. WaterSense has issued a notification of intent (NOI) to develop a specification for high-efficiency showerheads. The NOI outlines water efficiency and performance criteria for showerheads, and the standards to be created will involve input from industry and water-efficiency experts.

Resources

For more information, visit *www.epa.gov/watersense*.

CHAPTER **14**

Plumbing 2

Tankless Water Heaters

14.1   Supply Meets Demand

The water heater in an average house has a storage tank that holds between 30 and 50 gallons of water. When cold water enters the house, from a municipal supply or a well, it is split in two parts. One goes directly to the cold taps in the plumbing fixtures, plus any toilets. The other goes to the water heater, which in turn supplies hot water to the house. A few dishwashers and clothes washers can use just cold water because they have their own heating elements, but all fixtures with hot water inlets draw from the water heater.

14.2   Saving Energy

Whether it uses a gas flame or an electric element for heating, a water heater with a tank consumes energy even when the hot water is not running. That's because the water in the tank eventually cools down, and needs to be reheated so it's ready to use. Insulation, both built-in and added on, can help keep its contents warm. Even so, this arrangement means a water heater with a tank is using energy all the time, whether or not you need it to.

One other large item in your house consumes energy all the time, even more often than your space heating and cooling systems: the refrigerator. Water heaters and refrigerators both use large amounts of energy. After space conditioning, they are the largest energy consumers in most households.

The difference in energy needs between a refrigerator and a water heater is that one has to run all the time in order to preserve food and

Photo courtesy of Eemax, Inc.

drinks for freshness and safety. The contents of your refrigerator have to be kept cool to avoid spoiling, while the contents of your water heater do not. When you need hot water, it doesn't matter whether it was heated last week or just a minute ago.

Viewed another way, the practice of keeping a tank full of hot water makes little sense. You may run the hot water for an hour a day to shower, bathe, wash hands, and wash dishes. On laundry day, you use more hot water, but still only a few hours out of the day. Yet the tank on your water heater consumes energy 24 hours a day, just to be ready for the few hours of supply you actually use.

By operating only when it's needed, a tankless water heater saves all the energy that is otherwise consumed keeping a reserve supply warm. That can be as much as half of the energy a water heater uses. Tankless water heaters also make more efficient use of fuel.

14.3   Energy Factor

One way to compare the performance of water heaters is to check their energy factor (EF) ratings. Energy factor ranks energy efficiency by the average amount of hot water produced for each unit of fuel consumed. This number is made up of cycling losses (loss of heat as water circulates through the tank and pipes), recovery efficiency (how efficiently heat is transferred to the water), and standby losses (the amount of heat lost each hour from a storage tank, compared to the heat content of the water).

Higher EF numbers translate to greater energy efficiency. When you are considering several choices, such as different fuel sources and tank options, the EF is a useful guide. It won't tell you how much a water heater will cost to operate, but it does provide a common reference. Comparing models that use natural gas or liquid propane, for example, water heaters with tanks range from EF 0.59 to 0.65, while tankless heaters get 0.69 to 0.86.

In general, gas-fired tankless water heaters provide more heat faster than electric models. Because the use of natural gas or propane involves combustion, though, you will need to meet code requirements for venting, combustion air, and gas lines. Most electric models use 110 to 120 volt or 220 to 240 volt current. If possible, an electric water heater should have its own circuit breaker on the service panel.

Specialty models may be available for use with fuel oil, solar power, or geothermal energy; most are made to work with natural gas, liquid propane, or electricity.

14.4   Saving Time

When the water in a storage tank cools down, you need to run the tap until the cool water is flushed from the pipes. And once you use as much water as the tank had stored, you're out of luck until it heats another tankful. By contrast, a tankless water heater has a startup time of just a few seconds, so barely any water passes through the system before hot water is available. If you have chosen the right size tankless heater for your needs, it will keep heating water on demand, as long as you want.

Newer models of tankless water heaters have internal thermostats and flow meters. They adjust the heat according to the volume of water being used, which provides consistent temperatures and uses only as much energy as required. When you turn off the hot water at the fixture, the tankless heater switches off automatically.

Tankless water heaters can save time over the long term as well. Water heaters with tanks are only expected to work for about 10 years. Only the most expensive models are warranted to last longer. The leading tankless water heaters available today promise useful life spans of 20 years or more. Once you reach the payback period, which will depend on the efficiency of your current water heater and the tankless model you choose, the remaining service time is pure savings.

14.5   Saving Space

Most of the bulk in a conventional water heater is the tank itself. Without a large storage tank, all you need is a heating element with a pipe running through it. Of course, tankless heaters also contain valves, control circuits, safety guards, and so on, but even the largest tankless heater is much more compact than the smallest model with a tank. Most can be mounted on walls, freeing up the floor space required by a heater with a tank.

Another side effect of storing water is that mineral deposits can form in the tank. Over the lifespan of the tank, these deposits can build up to critical levels, then reduce the energy efficiency of the heater, or even cause the tank to corrode and leak.

14.6   Choosing the Right Model

If you choose a water heater with a tank, you should probably buy the largest one available. That sounds like strange advice, but it makes sense. Because larger tanks hold more water, the greater mass of the tank's contents will help maintain the temperature. More importantly, larger tanks often have thicker insulation and longer warranties.

For a tankless model, consider how much hot water you need each day, how much you need at any one time, and whether the water heater is for part of the house or all of it. According to manufacturers, tankless water heaters are about 30 percent more efficient than water heaters with tanks in households that use no more than 40 gallons of hot water each day. Households that use twice as much water will get about half as much improvement.

Resources

To learn more about tankless water heaters, look for information from manufacturers such as Bosch, Bradford, Chronomite, Eemax, Rheem, and Stiebel. For general information on water heaters, visit The U.S. Department of Energy Web site at *http://www.eere.energy.gov*.

CHAPTER 15

Heating

Increase comfort while reducing costs

Radiant In-floor Heat

15.1   No More Cold Feet

By now, you know that space conditioning—heating or cooling living
areas—accounts for about half of your household energy costs. Insulating and weatherproofing will help you control the indoor climate,
but you still have to decide how to provide heat and cooling.

One method of heating that has become popular in recent years is
radiant in-floor heat. It involves either wire mesh that converts
electricity into heat, or flexible tubing through which hot water is
circulated. Both versions provide easy, invisible comfort. They have
different requirements, though, so consider the details carefully
before you decide which to use.

15.2   Traditional Space Conditioning

Early furnaces and boilers used gravity to help circulate heated air
and water through a house; later models added fans and pumps to
improve range and control. Both methods involved large, bulky, complex conduits: ducts and registers for air heat, or pipes and radiators
for hot water. Air supply and return registers take up little room in
your living spaces, but the ducts shape the way the entire inside of
your house is put together. And pipes that run to and from hot water
radiators take up less space than air ducts, but the radiators are large,
heavy, obtrusive, and generally immovable.

These traditional methods of heating have other drawbacks as
well. Forced-air systems create excess pressure inside the house,

65

Photo courtesy of Plastic Pipe and Fittings Association

which increases the flow of indoor air to the outside. This contributes to heat loss in the winter and reduced cooling efficiency in the summer. Moving air also keeps dust, pollen, and other irritants in circulation, aggravating allergies and asthma.

Hot water radiators will not disturb the air, which makes them better suited for people with respiratory ailments. But they can't be used for cooling, and are often part of an older boiler system that may no longer be efficient. Hot water radiators can be moved if the entire system is shut down and drained, the pipes are disconnected and reconfigured, new holes are drilled in floors and walls, the system refilled, and air bled from all the radiators. For practical purposes, hot water radiators are an all-or-nothing proposition.

15.3 In-floor Heating Systems

Electric floor heat has several features that make it an appealing do-it-yourself project. First, it only requires standard 110 to 120 volt household current. Some kits consume as little as 120 watts, about the same as two incandescent living-room lights. The wire mesh can be attached directly to a wood subfloor, or to concrete with special fasteners. As soon as the mesh is in place, you can install the main flooring surface right on top of it. If you only plan to add radiant heat to one room, this approach is the simplest.

Radiant heating that uses hot water is called a hydronic system. Installation is more involved than for electric floor heat. Instead of a wire mesh in a fixed size that is simply spread out to cover an area, a hydronic system requires tubes to be routed in a way that covers the area without overlapping. To hold the tubing in place, special brackets are fastened to the subfloor. The tubing then has to run to and from heating and regulating equipment. This scale of project is usually, though not always, more appropriate for new construction.

15.4   New Construction

When you are building an entirely new floor, you can plan for radiant floor heating with few compromises. The area will be open and available for a thorough installation without shortcuts or deviations. This provides the best setting for a hydronic system. With full access, you can route the tubing exactly where you want it to go, and create the framing necessary for pouring a gypsum cement or concrete floor around the tubing. Then, if you install ceramic tile as the top surface, your floor will be strong, stable, and warm.

Hydronic systems offer some advantages over electric ones. They can be powered by a variety of energy sources, which can make a significant difference if electricity is expensive in your area. To provide the hot water needed for a hydronic system, you can use a water heater that runs on natural gas or liquid propane, a boiler that burns wood or runs on electricity, or other sources such as solar collectors, heat pumps, or geothermal energy.

Once the water is heated, it circulates quickly through the tubing—but you won't hear it the way you can a forced-air furnace or even a hot-water pump. It's almost silent. And because the water retains heat for a while after the heat source switches off, it cools down gradually, not suddenly. Also, because the heat comes from below, you can have the thermostat set a few degrees lower than usual but feel just as warm as before.

The trade-off for radiant floor heating, especially hydronic systems, is between the initial investment and the benefits that follow. A hydronic system will cost more per square foot than most other heating methods. But once installed, it will provide greater comfort and reliability than any other source of indoor heat. And hydronic systems are as much as 30 percent more efficient than forced-air systems for heating the same amount of space.

15.5   Remodeling

If you are renovating or adding on, and your project includes creating or removing and reinstalling a floor, an electric radiant system is easy to add. Once you decide what flooring surface you want, check to see if it is compatible with electric heating mesh. In general, electric radiant heating works with floating floors, most laminates, and vinyl. Two cautions: if the top flooring surface requires nailing, don't use electric mesh because you could easily damage the mesh underneath; and heated asphalt felt paper smells terrible, so use rosin paper as part of the underlayment instead.

In addition to the electric heating mesh, a floor-warming system has a hard-wired connection to the household electrical supply, a thermostat and a timer. The thermostat lets you control the floor temperature. Depending on the model of timer you get, it can switch off the heating system automatically after a specified period, or turn it on and off at predetermined times.

This method of heating has its limits. To prevent damage to the flooring surface, the mesh should not exceed safe temperatures for the chosen material. In some installations, that means the radiant floor heating will not be enough to warm the entire room. The amount of current you need to run the system will depend on its size. If you want or need to create a separate circuit for radiant floor heating, either be sure you can work safely with the electrical service panel, or hire an electrician to help you.

The resistance wires that provide the heat must not be bent or cut. Few rooms provide neat rectangles, though, so most manufacturers will create a mesh to your specifications. Make sure you have this option before you choose. During installation, check the resistance often to be sure the wires have not been damaged.

15.6   Retrofitting

There's one way to add radiant floor heating without tearing up the old floor or building a new one. If you have access to the underside of the subfloor, as you might in the basement under the main floor, you can add a hydronic system from below. You can find special aluminum plates, designed to be fastened under the floor with self-tapping screws. These plates have integral channels to hold hydronic tubing, which can be routed to a source of hot water. Because the temperature of a single-loop circuit like this has to be carefully controlled, you will need to connect it to a separate boiler, or to a dedicated zone on a boiler system with an appropriate manifold.

Resources

To learn more about radiant floor heating, visit *www.energystar.gov.*

CHAPTER 16

Ventilating

Reducing fuel costs and emissions

Heat and Energy Recovery Ventilators

16.1   Keeping in the Heat

Every house that uses fuel for heating loses some of its heat through exhaust airflows. A heat recovery ventilator (HRV) uses a heat exchanger to heat or cool incoming fresh air, which can reduce the energy consumption of a ventilating system by more than half. This makes effective use of heat energy that would usually be wasted.

An HRV that also exchanges moisture between indoor and outdoor air is called an energy recovery ventilator (ERV). This technology takes moisture from incoming humid air and puts it in the exhaust air to reduce the humidity level indoors. An ERV is best suited for use in climates with both high heat and high humidity.

Both HRVs and ERVs use fans to circulate indoor air. They create a slow but steady flow of fresh air into the heating, ventilation, and air conditioning (HVAC) system, and draw stale exhaust air out. An HRV or ERV with one exhaust fan relies on displacement to bring in a supply of fresh air; a system with two fans is called balanced, and the fans are meant to provide equal flow on exhaust and intake.

16.2   How an HRV Works

At the core of each HRV is a heat exchanger, which is made of materials that conduct heat quickly and efficiently. The heat exchanger has two ducts—one for incoming fresh air, the other for outgoing exhaust air.

Photo courtesy of Honeywell

The two ducts remain separate, so while heat is transferred from one to the other, the air streams do not mix. Instead, heat from the exhaust air is used to raise or lower the temperature of the incoming air.

The incoming air, once it has been heated or cooled, is routed to the existing HVAC system. There, it can be filtered and further heated or cooled, then mixed with other indoor air to provide a steady flow through the house. Most experts recommend a complete change of room air every three hours. With the help of an HRV, this can be accomplished with lower fuel costs because the fresh air requires less heating or cooling.

16.3   Differences between HRV and ERV

An HRV uses a recovery core that exchanges heat between incoming and outgoing airflows. The core is usually made of several aluminum or plastic plates. An ERV also uses a series of desiccants or permeable plates to transfer moisture between the two airflows. In other words, an ERV works in the same way as an HRV, except that it also transfers moisture from the incoming air to the outgoing air.

This helps control the humidity of indoor air, which can have as great an effect on personal comfort as the temperature. While ERVs are most often recommended for use in hot, humid locations during warmer months, they can also help prevent dry indoor air in cold climates during the winter.

16.4   Benefits in Comfort and Health

A regular flow of fresh air helps make a house more comfortable—and healthier. Newer houses, built to be virtually airtight, need active moisture control to prevent the growth of mildew and mold. These organisms can spread throughout a house and contribute to respiratory problems for the people in it. Opening a window or running an exhaust fan in the kitchen or bathroom will help circulate air for a moment, but these do little to manage the indoor climate over the long term.

16.5   Installation Options

Heat recovery ventilators depend on airflow, so they are most often used with forced-air systems, not hydronic or radiant systems. While it's possible to add an HRV to a house without a forced-air system, the HRV then needs one or more fans to circulate incoming and outgoing air. It must also depend on the difference between denser, cooler air and warmer, less-dense air—just like the gravity furnaces of a century ago.

Because it is designed to affect all the air that flows through a system, an HRV is most often installed as a whole-house unit. Some models of HRV or ERV are made for placement in a window or on a wall, but these are best for specific applications that are restricted to one room or area. For example, the exhaust fan from a bathroom or kitchen may also draw significant amounts of heat out of the room, and therefore the house. If the main HRV system is too far removed from one of these locations and the heat loss is great enough to require one, a separate HRV may be worth installing.

16.6   Reasons to Use an HRV

In new construction and in remodeling, one goal for builders is to create a "tight envelope." That means sealing the indoor space effectively from all outside elements. Many products mentioned in this book, such as spray-foam insulation and structural insulated panels, are valued for their ability to keep spaces isolated from unplanned airflow.

A tight envelope is desirable because it gives you greater control over the indoor climate, keeping heat in and moisture out—if that's where they are already. But if you need to change the temperature or humidity of an indoor space and it has no natural airflow, you need mechanical ventilation. In fact, building codes require it for most new houses.

As long as you need to have a method of cycling your indoor air, an HRV or ERV brings numerous benefits for little extra money. Adding an HRV to a new forced-air system will increase the cost by

$2000 to $3000; working one into an existing system, of course, will require more time, effort and expense. If you're building a new house and are not ready for an HRV yet, you can have the ductwork roughed in for use later on.

16.7   Finding the Right Model

The models of HRV and ERV units available to you will depend on the brands carried by your local distributors. And each model may offer a range of configurations, which can make choosing more difficult. For example, the same basic models may offer optional air filters, speed controls, or air quality sensors; in colder climates, they may also have defrost controls or preheaters.

To compare the specifications of different models, get information from the manufacturers—but be sure to look for independent test results as well. The Home Ventilation Institute (HVI) provides performance ratings on HRV and ERV systems, including airflow capacity, recovery efficiency, and sound levels.

Before you decide what kind of HRV to install, talk with HVAC contractors in your area. They should know which systems are appropriate for the climate you live in, and whether the added control over moisture makes an ERV a good investment. Another source of information is a distributor of HRV equipment, who will have broader knowledge of the industry, and may be able to recommend the best qualified installers. Also, your local building inspections department will have an HVAC specialist, whose advice can be both valuable and free.

16.8   Before You Install an HRV

Professional HVAC installers know what will affect the success of a new system. They recommend you take the following factors into consideration as you make your plan:

- Install the fan component in an accessible location for easier cleaning and service
- Make sure the fresh air intake is well away from chimneys, exhaust vents, and driveways
- Provide a separate supply inlet for each bedroom, plus one for each shared space such as a living room or study
- Provide a separate return outlet in each room with high humidity, such as a bathroom, kitchen, or laundry room
- Place each return outlet near the ceiling, and well away from any range or cooktop
- As with any air-handling system, use smooth, round ducting and the shortest routes possible

- Where ducts pass through uninsulated spaces, insulate the ducts themselves
- Seal any joints where ducts meet plenums, tees, wyes, registers, and so on
- If the system is likely to collect condensation, install a drain to draw moisture away from the heat exchanger

Resources

For more information, visit the Home Ventilation Institute (HVI) Web site at *www.hvi.org* or the U.S. Department of Energy Web site for energy efficiency and renewable energy at *www.eere.energy.gov* and search for "energy recovery ventilation systems."

CHAPTER 17

Air Conditioning

Improving indoor air quality and saving energy

Whole-House Fans

17.1 Keeping Your Cool

A well-designed ventilation system helps you control indoor temperature and humidity, as you have read in Chap 16. With a heat recovery ventilator (HRV) or energy recovery ventilator (ERV) unit added, it can reduce your heating costs and improve overall health and comfort. But as you read in the last chapter, most HRV and ERV units work best with forced-air heating, ventilation, and air conditioning (HVAC) systems. They provide less of a benefit in houses with hydronic or radiant heating.

Houses without forced-air systems generally rely on ceiling fans and window air conditioners for cooling in the warmer months. Unfortunately, ceiling fans only affect the air in one room—and while window air conditioners can cool and dry more air, they are generally expensive to run. This is where a whole-house fan makes a huge difference.

17.2 Energy Efficiency

During warmer months, a well-sealed house holds in heat. As soon as the temperature outdoors is higher than it is inside, the house begins to store heat, and stays warm even after outdoor temperatures go down. Outdoor temperatures are generally cooler in the evening, at night, and in the early morning. These are the best times to draw in cooler air to replace stale, heated indoor air.

Depending on the difference between indoor and outdoor temperatures, and on your personal preferences, a whole-house fan

iStockPhoto

may be all you need to cool your house even at the height of summer. In the hottest climates, it can supplement your air conditioning to give you better cooling at lower cost.

The chief benefit of using a whole-house fan is that it provides cooling for large areas at relatively little expense. The initial cost of a whole-house fan can be as little as $150; models with automatic features, higher flow rates, and advanced noise reduction cost more, but are worth a look. Air conditioners cost at least $250 for small window-mounted models, and central air conditioners start in the thousands.

Operating costs for whole-house fans are much lower as well. Comparative studies of the two cooling methods have shown that air conditioners cost 4 to 20 times as much per hour to run as whole-house fans. Granted, whole-house fans do not remove moisture the way air conditioners do, but the difference in operating cost remains significant. And whole-house fans may be better for your health.

17.3   Health Considerations

If you're like most people, you spend most of your time indoors. And although you may notice outdoor airborne pollution such as smog, exhaust fumes, smoke, and so on, indoor air is more likely to contain substances that can harm your health. For example, the following pollutants are much more common indoors, and in greater concentrations than outside:

- Asbestos
- Combustion gases
- Formaldehyde
- Lead
- Pesticides
- Radon
- Tobacco smoke

A publication from the U.S. Environmental Protection Agency (EPA) (*www.epa.gov/iaq/pubs/insidest.html*) includes this warning:

"Air inside our homes is up to 100 times more polluted than outdoor air. If too little outdoor air enters a dwelling, microorganisms can collect to levels that pose health and comfort problems. This problem can be especially serious in the energy efficient homes built in the last twenty years. These homes may not allow enough air changes, keeping contaminated air in and fresh air out."

In fact, the EPA lists indoor air quality as one of the five most urgent environmental risks to public health. Other organizations, including the National Academy of Sciences Institute of Medicine, also cite indoor air quality as a leading public health concern. While you can reduce the number and volume of these pollutants, the best way to handle them is to keep the air moving.

17.4   How Air Flows through a House

Older houses are generally not well sealed. That's partly because traditional building techniques and materials were less airtight, and partly because of the way structures age. In warmer climates, where heating concerns have been less of a priority for homeowners, houses can be almost porous.

Houses like these allow outdoor air to enter in several ways. One kind of airflow is called infiltration. That's where outdoor air enters the house through intentional and accidental openings, such as joints between walls, floors and ceilings, and around the edges of doors and windows. Another kind of airflow is called natural ventilation, which just means the movement of air through windows and doors as a result of wind, or differences in temperature and pressure. The third kind of airflow is mechanical ventilation—forced-air systems, air conditioners, exhaust fans, and so on.

If your house has many air leaks, you get less control over the indoor environment. But if your house is well sealed against air leaks, not much indoor air goes out, and not much outdoor air comes in, which results in stale, uncomfortable, and potentially unhealthy air. The ideal balance is a "tight envelope" with a system in place to circulate the air on purpose.

17.5   Calculating Air Exchange Rates

The amount of time it takes to replace indoor air with outdoor air is called the air exchange rate. The Home Ventilation Institute (HVI) recommends installing a whole-house fan, also called a comfort ventilator, powerful enough to replace the air in any room within two minutes.

A simple formula for calculating this airflow is to multiply the square footage of your house, including unoccupied areas, by the height of the ceiling. In this example, that measurement is eight feet. Choose a fan that moves at least half that figure in cubic feet per minute at 0.1 inches static pressure. Here are three calculations using this formula:

- 1000 square feet × 8 feet = 8000 × 0.5 = 4000 cubic feet per minute (cfm)
- 2000 square feet × 8 feet = 16000 × 0.5 = 8000 cfm
- 3000 square feet × 8 feet = 24,000 × 0.5 = 12,000 cfm

As you can see, larger houses require much higher volumes of air to be moved. Some whole-house fans are available in these greater capacities, but their cost is generally higher as well. You can also supplement a smaller whole-house fan by using your existing ceiling fans or an oscillating fan in each room.

Another consideration is where the indoor air will go. If you exhaust a whole-house fan into an attic space rather than directly outside, you will need somewhere for the air to go until it disperses through roof, gable, and soffit vents. To find the exhaust area you need in square feet, divide the capacity of your whole-house fan by 750. For example:

- 1000 cfm ÷ 750 = 1.50 square feet
- 2000 cfm ÷ 750 = 2.67 square feet
- 3000 cfm ÷ 750 = 4.00 square feet

If your whole-house fan vents directly outside, these figures will not matter. In fact, a powerful fan can create enough negative pressure within a house to make it uncomfortable. Be sure you have an adequate supply of incoming fresh air before you switch the fan on.

17.6   Features and Options

Traditional whole-house fans operate at one speed, and are controlled by a single switch. Newer models offer a range of choices, from multiple speeds to automatic controls and motorized dampers.

A fan with optional high-speed operation will let you cool a heated house quickly, then turn down the speed for normal circulation. A timer or thermostatic switch will let you choose whether to let the fan switch on or off by itself. And motorized dampers help prevent air from flowing back through the fan when it is idle. All of these are impressive, but remember to choose the features you believe you will really use.

Resources

To learn more about indoor air quality, visit the Home Ventilation Institute Web site at *www.hvi.org* or the U.S. Department of Energy Web site for Energy Efficiency and Renewable Energy at *www.eere. energy.gov* and search for "whole house fans."

CHAPTER 18

Insulating

Controlling Your Indoor Environment

Spray-Foam Insulation

18.1 Benefits beyond Warmth

By now, you have a good idea of how important ventilation is to your comfort and health. And in order to have control over ventilation, you need to seal your house against unwanted airflows. That means adding or improving your insulation.

The financial and environmental benefits of insulation are clear. If more than half of the money you spend on energy goes to space conditioning, as it does for most of us, any amount you can save there will have great impact. And by reducing your consumption of heating fuel, you both conserve resources and cut down on emissions from combustion.

Insulation does more than just keep indoors warmer in winter and cooler in summer, though. Combined with a thoughtfully designed ventilation system, it helps you manage the flow of air, moisture, and heat between the indoors and the outside. That includes exhausting stale air, filtering out allergens, preventing mold and mildew, and maintaining a comfortable balance of temperature and humidity.

In other words, the real value of insulation goes beyond savings on heating costs; insulating products that seal air leaks effectively provide the greatest number and degree of benefits.

18.2 Other Properties of Insulating Materials

All insulation should help keep heat on one side of a wall and cold on the other. Beyond that, it can have several other characteristics, which

Photo courtesy of Icynene, Inc.

you should bear in mind as you consider different materials. Here are the most common features mentioned for insulation:

- *R*-value—resistance to heat flow; a reference for comparing different materials
- Resistance to fire—marginally increases safety within the insulated spaces
- Moisture control—determines effectiveness at preventing water damage and mold
- Weight—affects ease of installation and likelihood of settling later
- Stability—provides estimated loss of insulating capacity from settling
- Convective heat loss—transfer of heat through solid materials, not air

18.3 Common Insulating Products

Most insulating products currently available fall into these categories:

- Structural insulated panels made of various materials (described in Chap. 1)
- Loose fill, such as recycled cellulose (described in Chap. 10)

- Rolls and batts, usually fiberglass and often with paper or foil facing
- Rigid foam boards, using polymers such as expanded polystyrene
- Radiant barriers, panels that help reduce convective heat loss
- Reflective insulation, which resists conduction, convection, and radiant heat transfer
- Insulating concrete forms, wall sections that combine concrete and insulation

All these products are effective, and any of them may be the best for your building or remodeling project. One other material, however, deserves special attention: spray-foam insulation, best known by the brand name Icynene.

18.4   Spray-Foam Insulation

The most visible difference between spray-foam insulation and other materials is the way it is installed. The foam begins as a liquid, and is sprayed on wall cavities, between joists, and in other open spaces with a high-pressure, low-volume applicator.

This liquid fills even the smallest gaps and contours in the receiving surface, and expands by 100 times its liquid volume to form a semi-rigid layer of light foam. As it expands, it continues to fill small spaces, sealing the area against air infiltration. When the foam dries, it can be trimmed flush to the studs or joists with just a handsaw.

By sealing the open spaces, spray-foam insulation reduces fuel consumption and costs, keeps out allergens and pollutants, and prevents some outside noise from entering the house. One thing it doesn't do is trap moisture.

18.5   Managing Indoor Air Quality

The spray-foam insulation known as Icynene has open cells, like a sponge. It's breathable. Because it minimizes air infiltration, though, it does not draw moisture from the outdoor air. When water gets in a space insulated with sprayed foam, the moisture runs through or evaporates freely. This reduces condensation, and the growth of mildew and mold.

This also means spray-foam insulation helps you focus on the intentional flow of air through your house, which gives you control over pollens and pollutants. Following the builder's motto of "seal it tight and vent it right," you can keep respiratory problems, such as asthma and allergies, in check.

Even during application, spray-foam insulation makes breathing easier. For example, Icynene is water-blown, and contains no HCFCs, HFAs, HFCs, HCs, formaldehyde, or VOCs. It does not cause corrosion, is the only insulation certified by the Envirodesic air quality improvement certification program, and meets energy efficiency standards set by ALA Health House, EarthCraft, and Energy Star.

Also, thanks to its efficiency at filling spaces, spray-foam insulation helps reduce noise entering the house, and even noise transmitted from one part of the house to another.

18.6   Saving Energy and Money

All insulation products, correctly installed, will help improve indoor comfort and reduce heating costs. Spray-foam insulation is particularly good at sealing the envelope, so it brings the added benefits of minimized air infiltration.

In new construction, spray-foam insulation can seal a house well enough for a smaller, more fuel-efficient boiler or furnace to provide all the heat the house needs. This may also help the homeowner qualify for favorable mortgage rates, rebates on heating equipment, or even tax advantages. The payback period is short, and, amount you save on fuel will only increase as energy prices rise. And, any house built for energy efficiency will both cost less to own and bring a higher resale price.

Remodeling projects that include the use of spray-foam insulation may require a different application method. Using Icynene as an example again, it is also available in a pourable version, which can be added to existing spaces without tearing out walls or floors. The liquid can be poured through small holes, but will expand to fill and seal large areas. This makes expanding foam an effective material for insulating irregular and hard-to-reach spaces.

18.7   Back to Basics

R-value is the one characteristic on which all insulating materials are judged. Numbers are given in value per inch of thickness. Spray-foam insulation has inherently high R-values, higher than fiberglass batts or loose fill and comparable to rigid foam sheets. But its sealing properties virtually eliminate airflow, so it is more effective than its R-value alone indicates.

The U.S. Department of Energy acknowledges the limits of information provided by R-value. Real-world conditions such as material compression or settling, air gaps, and conduction through adjacent materials—including the building's framing members—all affect the performance of insulation.

Perhaps a better measure is actual energy costs after installation. A house sealed with spray-foam insulation costs less to heat than one with batts, loose fill, or rigid foam sheets. Estimates vary from 10 to 50 percent, but one thing is constant: a house with spray-foam insulation always saves more on heating.

Resources

To learn more about insulation in general, visit *www.eere.energy.gov*; for information about Icynene, visit *www.icynene.com*.

CHAPTER **19**

Cooking

Harnessing the power of magnetism

Induction Cooktops

19.1   New Methods for Old Chores

In some ways, preparing food on an induction cooktop is unlike any
other way of cooking. Instead of heating a gas burner or electric
resistance element, magnetic induction cooking uses electricity to
produce a magnetic field. Within this field, iron atoms react to electric current by vibrating at high frequencies. The resulting friction
causes the object containing the iron—in this case, the pan—to heat
up quickly.

In other ways, however, cooking with an induction cooktop is
pretty much the same as using a traditional range. You still put
ingredients in a pan and use heat to help mix their flavors and change
their consistency. By itself, having a cool cooktop surface while the
pans are hot is little more than a novelty. The greatest benefits are in
using the device, rather than in the results it produces.

19.2   Understanding Induction Cooking

A traditional electric cooktop uses a coil through which electricity
passes. The coil is designed to convert electricity into heat. Like the
filament in a light bulb or the element in a toaster, the coil can't handle the entire current load, so it gives off excess energy in another
form.

An induction cooktop also uses an electric coil, but a different
kind. This coil converts the electric current into a high-frequency
electromagnetic field, to which ferrous materials react as if they are
heating elements. Instead of shedding excess current, though, heat

Photo by Clayton Bennett

results from vibration of the atoms within those magnetic objects. Strictly speaking, electricity is passing through the pan, and resistance within the magnetic material generates heat. If there's no ferrous material in the magnetic field, nothing gets hot.

19.3   The Biggest Differences

Cooks who are fond of natural gas ranges often mention the features they like best about it, at least compared to electric resistance elements. A gas flame is easier to start and stop quickly, and the cook can judge the amount of heat generated by the height of the flame. Those who prefer electric resistance cooktops mention their consistent performance and safe fuel source, which is also a convenience in locations without natural gas service.

Both gas and electric resistance cooktops provide a hot surface, and the user places a pan on top of it to transfer heat from the cooktop, through the pan, to the food. An induction cooktop does not get hot by itself, but causes the pan to heat up. The pan itself becomes the heating element. This has several benefits; the most obvious is energy savings.

Because little heat is lost to the surface underneath the pan or to the air around it, the food is heated more quickly, and with less energy consumed. This transfer of energy from electricity to heat through induction is much more direct than either electric resistance elements or gas flames. Manufacturers estimate the energy efficiency of induction cooktops at 80 to 90 percent—far better than the 55 to 65 percent efficiency of electric resistance or gas cooktops.

19.4   Additional Capabilities

Underneath the glassy ceramic surface of an induction cooktop, a set of elements convert electric current into a magnetic field. The heat results when a cooking pan with magnetic properties is placed in that field. This means only ferric pans—those containing iron—will work with induction cooktops. Many of your current cooking pots and pans may still work, but those made of copper, tempered glass, aluminum, and other materials will not.

When there's no magnetic material on the surface, it will be cool to the touch, even when it's switched on. An induction cooktop only gets hot when heat is transferred from a pan to the cooktop. Because the pan and its contents are hot, you will still need to use care, but your risk of accidental burns will be lower.

Along with more direct heating comes greater control. Some induction cooktops include sensors that detect the size of pan being used, and can adjust the power as needed. The result is greater precision through finer adjustments, something any cook will appreciate.

19.5   Potential Drawbacks

Despite its efficient operation, an induction cooktop has a long payback period. That's because most models cost three to four times as much as traditional electric resistance ranges or gas cooktops. As induction cooking becomes more popular, more manufacturers will offer these products, which should drive quality up and prices down. For the moment, though, the initial investment is high.

The other most likely problem with induction cooking is adjusting habits. New users have to set aside nonmagnetic pans, learn how unfamiliar controls affect cooking times, and remember to keep some items away from the cooktop. Aluminum foil, for example, will bond to the surface if it is in the magnetic field when the current is on.

19.6   Definite Advantages

An induction cooktop is as easy to install as one that uses electric resistance, and easier than one that uses natural gas. Models are powered by either 110 to 120 VAC household current or 220 to 240 VAC current, which is common for electric ranges, dryers, and some air conditioners. If you need to add a new outlet for an induction cooktop, make sure it has its own circuit—and hire an electrician if you're not confident about working directly in the service panel. Once you have a power supply in place, induction cooking is plug-and-play.

Although they are catching on, induction cooktops are still more common in other countries than in the United States. If you can't find the model you want through a domestic distributor, you may be able to import one yourself. Of course, an imported model may not have certification from a third party such as Underwriters Laboratories. Check its specifications with your local building inspector to make sure the cooktop and your installation plans will meet code requirements.

Cooks who are accustomed to using gas ranges appreciate the ability to control the temperature quickly. They also value the high output of higher-end models for professionals and serious amateur chefs. Induction cooktops offer equally fast changes in temperature, and comparable heating properties. Instead of measuring output in therms, however, the power of an induction cooktop is rated in watts. Natural gas also has some drawbacks, such as carbon monoxide and related combustion gases, plus potential health risks and explosion hazards; induction cooking has none of these.

For now, induction cooktops may seem strange to traditional cooks. But just like microwave ovens, they provide results cooks want, and will only grow in popularity.

Resources

To learn more about induction cooktops, visit *www.toolbase. org/ Technology-Inventory/Appliances/induction-cooktops* or *www.induction-cooktop.com.*

CHAPTER **20**

Washing

Save water, time, money, and clothes

Horizontal Axis Washers and Dryers

20.1 Getting More from Less

Energy-efficient technologies are designed to help you reduce your consumption of limited resources. This also means they save you money over the long term, but the initial investment is often higher. The time required for you to realize a financial benefit is called the payback period. Horizontal axis washers and dryers, more often called front loaders, can save enough water and energy within a few years to justify the additional cost. Another kind of payback, however, takes place almost immediately.

Front loading washing machines can handle larger loads than many top loaders, but use less water and electricity. And even while you earn back the extra money spent on a front loader, it gives you the same results or better, but more quickly. Their spin cycles also remove more water from freshly washed clothes, which means they need less time in the dryer—indirect savings of energy and time.

20.2 How Front Loaders Work

The axis of a traditional washing machine is vertical—a cylindrical drum with an agitator in the middle. The agitator is necessary because the clothes would not circulate from top to bottom without its help. Because a front loading machine uses gravity to keep the clothes moving, it doesn't need an agitator. This saves room in the machine and allows you to run larger loads.

Gravity also lets a horizontal axis washer make more efficient use of water. Instead of filling the entire drum, as a top loader does, it

Photo by Clayton Bennett

cycles the clothes through water in the lower half of the machine. Because a front loader uses less water, it also uses less energy to heat the water. When the wash and rinse stages are complete, the drum spins at high speed to force out as much water as possible—generally more than a top loader can extract.

The water and energy savings are substantial. According to figures from the U.S. Environmental Protection Agency (EPA), horizontal axis washers use 40 percent less water than top loaders, and as much as 50 percent less energy.

20.3 Increasing Availability

Until a few years ago, horizontal axis washers were difficult to find in the United States. Homeowners who were determined to buy them found they had to import their own, or buy them through distributors at high cost. The sales volume was too low for major retailers to carry them. One reason is the initial cost of front loaders. The least expensive models, which meant those with the lowest capacity and fewest features, had prices comparable to the most well-appointed and expensive top loaders. The scarcity of front loaders made parts and qualified repair service very difficult to find.

In the past decade, three things have changed to make horizontal axis washers more widely available. The first has been a steady increase in the costs of providing municipal water and electricity. The second is a consequence of the first: greater awareness of water and energy consumption. Consumers, municipalities, and manufacturers have all learned that conservation is the key to managing the future

costs of our essential utilities. The third is enthusiastic recommendations from owners of front loaders. These personal endorsements
have inspired more consumers to consider new technologies than
any promotion or rebate ever could.

Homeowners who manage their own appliances will find front
loaders just as easy to install as traditional top loaders. They have the
same kinds of water supply inlets and discharge hoses, and most use
110 to 120 VAC household current.

20.4   Benefits of Horizontal Axis Washers

By reducing your consumption of water and energy, a front loader
saves you money. Publications from the EPA estimate that a family of
four saves up to $100 per year in utility costs. A larger family, or an
especially active one whose clothes need washing more often, will
realize even greater savings. Reduced needs for detergent and drying
time help as well.

The U.S. Department of Energy conducted a field study of
103 horizontal axis washers, called the Bern Clothes Washer Study.
The results showed that front loaders used 56 percent less energy
and 38 percent less water, on average, than top loaders.

Government information on horizontal axis washers estimate
savings of $550 in operating costs over the life of the appliance
compared to a top loader—but this figure is conservative. Reduced
consumption of water, energy, and detergent for washing, plus
reduced energy consumption for drying, would provide that much
saving in five years. If the washer lasts longer than five years, or the
prices of water, energy, or detergent go up, the savings will be
greater still.

20.5   Choosing a Horizontal Axis Washer

All large retailers that sell major appliances now offer horizontal axis
washers. They generally cost more than traditional models, but as
you know now, the difference is paid back over time. Aside from
price, almost everything is the same as for top loaders. For example,
warranties vary by manufacturer, but most cover the entire washer
for one to three years, and some cover individual parts, such as the
motor or drum, for longer periods.

By law, all new washing machines must display a yellow-and-
black EnergyGuide label. The U.S. Federal Trade Commission requires
all manufacturers to provide energy use data, which helps you
calculate operating costs and compare different models.

Remember that EnergyGuide and Energy Star are not the same
thing. Energy Star is a program of the EPA that sets guidelines for
efficiency and performance. All major appliances must have

EnergyGuide labels, but not all are "qualified" for Energy Star status. A few other things to keep in mind:

- Energy Star qualified washers range from 1.6 to 3.8 cubic feet in capacity; instead of buying the largest model you can, choose one that meets your needs.

- Horizontal axis washers are made as stand-alone appliances, in stacked combinations with matching dryers, and even for installation under a counter.

- Some models allow you to adjust water levels to fit the load. If water conservation is already a high priority for you, this feature will be worth looking for.

- Modified Energy Factor (MEF) and Water Factor (WF) are important measurements; higher MEF values are preferable, while lower WF values are better. That's because MEF measures efficiency while WF measures consumption.

Finally, although most retailers carry horizontal axis washers, no single seller offers more than a few models. This makes true side-by-side comparisons more difficult. However, the EPA Energy Star web site offers lists of models, manufacturers and retailers to help you find the best front loader for your needs.

20.6   Choosing a Clothes Dryer

Washing machines are more complicated and have more options than dryers. Case in point: even though some models of clothes dryer are designed for energy efficiency, none of them receives Energy Star qualification. That's because most of them use roughly the same amount of energy. The best single feature a dryer can have to help you conserve energy is a moisture sensor. Using this sensor instead of a timer, you can have the dryer stop running as soon as the clothes are dry enough. This saves energy costs and prevents unnecessary wear to clothes.

Resources

For more information on horizontal axis washers, the Energy Star program, and even advice on getting better results from your laundry appliances, visit *www.energystar.gov*.

Lifestyle Changes.

To Change Your Life, Change Your Habits

As individuals and households, we make ecological decisions every day. The goods we consume and the waste we produce affect our immediate environments, and the combined result of our choices affects everything from our physical health to parts of the natural world we may never see.

Part 1 and Part 2 of this book discussed physical changes to features and systems in your house. Each of the materials and technologies described can help you improve your quality of life and reduce your impact on the natural world. Part 3 goes in a different direction. The following 10 chapters will describe changes, large and small, that you can make to the way you conduct your daily life.

Some of these changes are simple and straightforward. All you need to do is make a few phone calls and fill in a few forms. Whether you specify alternative sources for your electric power, take out an energy-efficient mortgage, or apply for tax credits to recover some of your expenses, your part is limited to initiating action. Other changes, such as growing your own vegetables or reusing graywater, require considerable effort and modest expense.

Managing Personal Consumption and Waste

No matter where you live and what you choose to do with your life, you need food and water. And, if you're like most people, you need a source of energy and a way to get rid of your food and water waste. The practices described in the following chapters assume you want to maintain or improve your quality of life, but are willing to take on some of the work necessary to conserve natural resources.

21. Personal Choices: Reducing Your Carbon Footprint
22. Energy Sources: Green Power
23. Financial Incentives: Saving Money Twice
24. Cooperative Buying: Community-Supported Agriculture
25. Agricultural Independence: Container Gardening
26. Eliminating Waste: Composting Biodegradables
27. Reducing Waste: Using Graywater
28. Washing Up: Making Your Own Cleaning Products
29. Second Chances: Reusing Containers
30. Getting Outside: Xeriscaping

A Word of Caution

If you are reluctant to take on any of these ideas, you probably won't enjoy the process and are less likely to get satisfying results. Don't try to change your habits for any reasons but your own. Only do these things if they make sense for the life you want to create.

CHAPTER 21

Personal Choices

The effects of buying decisions you make every day

Reducing Your Carbon Footprint

21.1   Changing Habits

No matter what you believe about the causes of global climate change, it's a fact of life on earth. The climate trends of the past hundred years may be the result of natural phenomena or entirely caused by human activities. Both factors are probably contributing, and we can't know for certain in what proportions. But some things are clear:

- Global climate change affects all living things.
- Weather patterns of all kinds are becoming more severe.
- Human activities contribute to global climate change.
- We can't affect the long-term processes of nature.
- We can learn more about the effects of our daily decisions.
- We can make decisions that reduce our carbon footprints.

This chapter will provide an explanation of the carbon footprint concept, outline some of the ways you can calculate your own impact on carbon emissions, and offer general guidelines for reducing them. Some of the suggestions may be impractical right now, but knowing more about the consequences of your choices will help you make improvements over time.

21.2   What "Carbon Footprint" Means

The term is used in slightly different ways, depending on the context. Here's a definition that works in most situations: Your carbon footprint

97

is the amount of carbon dioxide (CO_2) released into the atmosphere as a result of the direct and indirect consumer choices you make. Some causes, such as driving a car with poor gas mileage, are obvious; others, like buying produce out of season that must be imported from far away, can be harder to recognize.

21.3   Consumption Affects Emissions

Carbon dioxide is a by-product of combustion, which means human civilization has created carbon dioxide emissions since the first people learned to use fire for heat and cooking. It also means that countries with relatively low levels of industrial development can also contribute high levels of carbon dioxide emissions. In some parts of less-developed countries, the air has a blue tint from coal burned by hundreds and thousands of individual households.

Greater mechanization, however, creates even more carbon dioxide emissions. The fuel used for manufacturing, transportation, and electric power for modern conveniences results in higher amounts of carbon dioxide generated per person. As more people around the world gain access to cars, air conditioners, refrigerators, and other machines that consume fuel directly or indirectly, the global production of carbon dioxide will continue to rise.

The parallels with population growth are striking. Countries that developed industrial economies sooner also saw improvements in quality of life, including health care. Death rates fell sharply; then, within a few decades, birth rates fell also. The driving force was technology, which means these countries are now the best positioned to reduce carbon emissions.

The reasoning is simple: countries that are focused on economic development give less attention to the undesirable effects of growth. A country with a growing economy and consumer culture may well ignore the downside of industrialization. Western countries did exactly that in the twentieth century. The good news is that anyone, in any country, can help reduce carbon emissions, and the benefits reach around the world.

21.4   Global Causes, Global Effects

The atmosphere does not stay in one place or belong to anyone. It covers the globe. What affects the atmosphere in one part of the world reaches all others. So reducing carbon dioxide emissions anywhere on Earth will help everywhere else. But as with any distributed problem, no single change in human activity will make all the difference needed.

Some people call this a "non-point-source" situation. In other words, millions of individual actions, taken without any intent to cause harm, contribute to a recognizable result. That also means millions of conscious, deliberate actions are required to offset the others. This is the most encouraging thing about the situation: every one of us can make a difference.

21.5   Past, Present, and Future

We can compare the current composition of the atmosphere with the way it used to be hundreds or thousands of years ago. By taking core samples from ice at the poles, scientists can measure the levels of carbon dioxide and other compounds accurately for different periods in history. This information helps us understand how much the climate is changing, and how quickly.

According to these measurements, current levels of carbon dioxide emissions are 40 percent higher than they were before the development of mass production and machine-powered transportation in the 1800s. While that's a significant increase, some of it might be natural and cyclical. But viewed over a longer term, the carbon dioxide levels we have now are the highest they have been in more than half a million years. Because the effects we're seeing today are negative, we have an obligation to reduce our part of this change.

21.6   Decisions, Not Sacrifices

When you're accustomed to consumption, any reduction can feel like deprivation. That makes any change uncomfortable. It's even harder to give up something you enjoy for some concept of greater good that may never touch your life. But many environmental improvements also bring immediate benefits—things that make your life noticeably better.

Consider the examples here a starting point. Adopt the ones that are easiest for you, and think about other ways to accomplish the same goals.

At home:

- Maximize your insulation and weather stripping.
- Switch from incandescent bulbs to compact fluorescents.
- Install a programmable thermostat and reduce the temperature when possible.
- Use a consumption meter to identify inefficient or unnecessary appliances.
- Replace water-wasting fixtures with low-flow models.

At work:

- Use lighting for work and safety only. Wherever you can, install motion-sensor switches.
- Shut your computer all the way down when you won't be using it for an hour or more.
- Print fewer documents, and print two pages to a side, use both sides of the paper, or both.
- Take a bus or train to work, ride a bike, or telecommute if that's an option.

On the road:

- Make sure your car is tuned up and maintained on a regular schedule.
- Keep your tires inflated to the recommended pressures; this improves safety, too.
- Adjust your driving habits to follow speed limits and maintain steady road speeds.
- Avoid unnecessary car trips, and combine errands as you can.
- For shorter distances, consider walking or riding a bike.

21.7   Everyone Wins

All the suggestions above require you to choose and act, changing ways you might still be doing things. There's also one thing you can do by asking someone else to change habits, and everyone involved will be pleased to accommodate you.

According to 41pounds, a non-profit organization, the average American receives 41 pounds of unwanted commercial mail ("junk mail") every year. They claim that the energy used to produce, deliver, and dispose of all that junk mail produces more greenhouse gas emissions than 2.8 million cars would create in the same time. For a fee, the group offers to have your name removed from as many mailing lists as possible, and will donate part of the fee to Carbonfund, another group dedicated to reducing carbon emissions.

You'll be spared the nuisance of discarding offers you don't want, your letter carrier and recycling collector will have less material to handle, and the marketers themselves will be grateful not to spend money on advertising that is unlikely ever to result in a sale. Everyone wins.

Resources

For more information on calculating and reducing your carbon footprint, visit Carbonfund at *www.carbonfund.org*, carbon footprint at *www.carbonfootprint.com* and 41pounds at *www.41pounds.org*.

CHAPTER 22

Energy Sources

Buying electricity from alternative utilities

Green Power

22.1 Discovering Your Choices

By now, you know that reducing your energy consumption through conservation will help cut down on emissions from your own house and from power plants. And you can probably do more, even without going off the grid to produce your own electricity. The same power utility you are using now may have "green power" options available.

22.2 Changes in Recent History

Traditionally, electric power has been a natural monopoly—a service that makes little sense to offer through more than one provider per market. When Consolidated Edison was laying miles of copper lines under the streets of New York in the late nineteenth century, no other company could have done the same. This was the case for most utilities, and consumers grew accustomed to having one provider for each utility service.

Two developments of the past 20 years have given consumers greater choice. Deregulation has opened the marketplace to competition, and power utilities have begun buying electricity from independent providers that use renewable sources such as solar power, wind power, hydro power, geothermal energy, and even biomass fuels. Consumers in many states can now specify what sources they want for their electric power, sometimes without even changing service providers.

Electricity generated by these alternative methods is best known as green power.

103

iStockPhoto

22.3   Market Deregulation

As members of a regulated industry, electric companies are subject to specific legislation that affects the way they do business. Whether competition is allowed in a given market depends on legislation, usually at the state level, rather than on decisions by individual power utilities. The most recent figures available from the U.S. Energy Information Administration show that 23 states and the District of Columbia have partially deregulated electric power, and allowed competition in that market. Consumers in those states can buy their electricity from other providers if they want. The list includes

- Arizona
- Connecticut
- Delaware
- District of Columbia
- Illinois
- Maine

- Maryland
- Massachusetts
- Michigan
- New Hampshire
- New Jersey
- New York

- Ohio
- Oregon
- Pennsylvania
- Rhode Island
- Texas
- Virginia

In markets where competition is allowed, local distribution companies still provide transmission and distribution. Because deregulation varies by legislation, and therefore by state, different terms will apply for each location. In Arkansas, Montana, Nevada, New Mexico, Oklahoma, and West Virginia, deregulation has been discussed and legislation has been drafted, but no action has been taken recently.

22.4   A Good Place to Start

If your state is among the 24 that allow competition in electric utilities, you can find out about your options through the Green Power Partnership web site from the U.S. Environmental Protection Agency (EPA) at *www.epa.gov/greenpower*. The site includes a publication titled *The Guide to Purchasing Green Power*, which was compiled by the EPA, the U.S. Department of Energy (DOE), the World Resources Institute, and the Center for Resource Solutions (CRS). It's a free download, and it explains how to buy green power and lists different green power products.

22.5   Same Utility, Different Source

Green power may be available to you even if your state doesn't yet allow competition among electric utilities. Even regulated businesses have options, which means you do, too. At last count, hundreds of utilities in 30 states offered ways for their customers to specify environment-friendly fuel sources for the electricity they buy. These alternatives, called green-pricing programs, provide a way for you to encourage the utility to buy more power from producers that use renewable energy resources.

In general, these options do cost more than regular service from the same electric utility. Before you balk at the idea of spending more money for the same product, though, do two things: find out if your utility offers green-pricing options and get a firm quote on the price difference. Then you can decide whether the cost premium is reasonable for the change it represents.

22.6   Renewable Energy Certificates

Even if your state does not allow competition and your electric utility does not offer alternative options, you can support green power another way—buying renewable energy certificates (RECs). Sometimes called tradable renewable certificates, green tags, or green energy certificates, RECs are commodities that can be bought and sold. They provide evidence that power was generated from a renewable source. Each REC represents one megawatt-hour of electricity, a little more than an average consumer uses in a month.

To buy RECs, you keep your existing electric utility, but pay for proof that someone else is getting green power. This lets you encourage the use of green power if it isn't yet available to you. In effect, you are subsidizing the development costs of producing green power. Instead of increasing costs to discourage emissions, as carbon credits do, RECs offset capital expenses for electric utilities to make green power a more attractive option.

Here's how the process works: One REC is issued for each megawatt-hour of electricity generated by a green power producer. Each REC has a different identification number so that it is only counted once. When the power is added to the electrical grid, the REC can be sold.

22.7 Evaluating Green Power Sources

The electricity that reaches your house will be the same, no matter the source. So you may wonder how to tell where your green power money goes. Independent third parties can provide verification to help you feel certain about your choices. For example, the voluntary certification program Green-e evaluates RECs, electricity products, and green power pricing structures for about 100 green power marketers in the United States. And the EcoPower certification program from the Environmental Resources Trust also certifies RECs for the renewable energy attributes of green power. It makes no claims about environmental benefits; those are implicit, and up to you and the provider to evaluate.

22.8 Buying Power for Buying Power

If you're concerned that individual homeowners won't provide enough of a market for green power, you can relax. Hundreds of large organizations, public and private, have chosen green power for their facilities. These include federal agencies, military and aerospace installations, national laboratories, the United States Postal Service, manufacturers of all sizes, high-technology companies, financial institutions, defense contractors, department stores, hotels and resorts, major sports teams and leagues, airports, grocery stores, car makers, movie studios, churches, international political groups, states and cities across the country, schools and universities—and electric power utilities.

Resources

For more information on green power, visit *www.epa.gov/greenpower*.

CHAPTER 23

Financial Incentives

Tax advantages and energy efficient mortgages

Saving Money Twice

23.1   Investing in Futures

By now, you know that energy efficiency saves you money. Even home improvements that require an initial investment will end up costing you less over time than keeping things the way they are. These improvements also make your house more valuable, which appeals to buyers and lenders—and the reductions you make in energy consumption help utility companies defer the costs of building new facilities.

In fact, both government agencies and private businesses, such as mortgage lenders and power utilities, offer incentives to help you make the transition from traditional technologies to ones that are more energy efficient. This chapter will provide a brief overview of two programs: tax credits and energy efficient mortgages (EEMs).

23.2   Federal Tax Credits

First, an important distinction: tax deductions allow you to reduce the amount of your declared taxable income, while tax credits let you take an amount off your total tax bill. In most situations, tax credits save you more—even if the dollar amount is the same.

The federal government offers tax credits for activities that benefit the economy. Now that energy conservation is seen as an important national priority, you are allowed to claim tax credits for several kinds of home improvements, such as adding insulation, replacing windows, and installing more efficient heating and cooling equipment.

Here is the current "credit for nonbusiness energy property" list of product types:

- Exterior windows and skylights
- Storm windows
- Exterior doors
- Storm doors
- Metal roofs
- Insulation
- Central air conditioning
- Air source heat pumps
- Geo-thermal heat pumps
- Gas, oil, or propane furnaces or hot water boilers
- Advanced main air circulating fans
- Gas, oil, or propane water heaters
- Electric heat pump water heaters
- Solar water heating
- Photovoltaic systems

The list also includes credits for hybrid gasoline-electric, diesel, battery-electric, alternative fuel, and fuel cell vehicles, and for some other applications of fuel-cell technologies.

23.3   Limitations and Conditions

These credits are all subject to conditions, from minimum values of result to limits on the amount you can claim. And you may have noticed that most of these credits apply to renovations and repairs rather than to new construction. But if you're planning to make any home improvements, especially ones that bring environmental benefits, take a look at the incentives available to you. One of them may make a project more feasible, or allow you to do more than your existing budget will allow.

New houses won't qualify for some of these credits, such as those for "eligible building envelope components" or "qualified energy properties." That means windows, doors, insulation, roofs, and HVAC and non-solar water heaters. But you may still be able to get tax credits for photovoltaics, solar water heating, and fuel cells for new home construction.

23.4   Energy Star and Tax Credits

The Energy Star program from the U.S. Environmental Protection Agency (EPA) provides guidelines for construction materials and appliances, as you may remember from elsewhere in this book. The program also includes a comprehensive set of requirements for an entire new house to earn an Energy Star rating.

To qualify, a new house must be at least 15 percent more energy efficient than one built to the 2004 International Residential Code (IRC). It must also include additional energy-saving features to become 20 to 30 percent more efficient than standard houses.

Almost any kind of house can meet these requirements, as long as it's no more than three stories high. It can be a single family, attached, or low-rise multifamily house; a manufactured house; a systems-built house, such as one with structural insulated panel (SIP), insulated concrete forms (ICF), or modular construction; a log house; a concrete house; or even an existing retrofitted house.

For the purposes of Energy Star ratings, the systems that make a house energy efficient include effective insulation, high-performance windows, tight construction and ducts, and efficient heating and cooling equipment. By using Energy Star qualified lighting fixtures, compact fluorescent bulbs, ventilation fans, and appliances, a new house can provide even greater comfort, savings, and environmental benefits.

Not all Energy Star qualified houses or products qualify for tax credits. Standards for products that merit tax credits are even stricter than those for earning Energy Star qualification. Both kinds, of course, will save you money over the long term.

23.5   Getting Over the Threshold

While it will eventually cost less to own, a house that features energy-efficient improvements is generally more expensive to buy. Even though environment-friendly materials and technologies only add about five percent to building costs, they have far greater value in the eyes of buyers and lenders. The reason is simple: any improvement that saves energy and expense now will save even more in the future.

Government agencies such as the U.S. Federal Housing Administration (FHA), U.S. Department of Housing and Urban Development (HUD), and U.S. Department of Energy (DOE) offer information on ways for homeowners and builders to make energy efficiency a priority. One tool they recommend is an Energy Efficient Mortgage (EEM).

23.6   Energy Efficient Mortgages

An EEM has strict requirements and specific advantages. Whether it's a conventional mortgage, FHA, or VA, any EEM is designed to make energy-efficient houses more affordable to buyers. This means the house has to meet several criteria for energy efficiency, and the buyer has to fulfill all the other requirements to take out the loan.

23.7   Conventional EEMs

Conventional EEMs can be offered by lenders who sell their loans to the Federal National Mortgage Association or the Federal Home Loan Mortgage Corporation, better known as Fannie Mae and Freddie Mac. These conventional mortgages are adjusted by the same amount in two directions. Once the estimated energy savings are calculated, that amount is added to the borrower's stated income for the purposes of the loan. A conventional EEM offered through Fannie Mae also adds the value of energy-efficient improvements to the value of the house itself.

23.8   FHA EEMs

An EEM from the FHA allows a lender to add the total cost of energy-efficient improvements to an existing, approved mortgage. The amount is limited by both a dollar figure and a percentage of the house's value. These EEMs are not affected by regular FHA loan limits, and do not require any additional down payment. They are available for both site-built (permanent) and manufactured (mobile) houses, especially those qualified for Energy Star ratings.

23.9   VA EEMs

An EEM from the U.S. Department of Veterans Affairs (VA) is offered to qualified military personnel, reservists, and veterans. It helps pay for energy improvements that will be made to an existing house, although it limits the amount that can be borrowed.

23.10   Conclusion

An EEM benefits everyone involved in buying or selling a house. Buyers get favorable rates on desirable houses; sellers make their houses accessible to more buyers and close more quickly; builders and remodelers increase the initial and resale value of a house; and lenders make sound investments in mortgages and the houses that serve as their collateral.

Resources

To learn more about tax credits for energy-efficient home improvements, visit the Tax Incentives Assistance Project (TIAP) web site at *www.energytaxincentives.org*, the EPA Energy Star web site at *www.energystar.gov*, and Federal Citizen Information Center web site at *www.pueblo.gsa.gov*.

For information on energy efficient mortgages, visit Fannie Mae at *www.fanniemae.com*, Freddie Mac at *www.freddiemac.com*, HUD at *www.hud.gov*, the FHA at *www.fha.com*, the VA at *www.va.gov*, or the Residential Energy Services Network (RESNET) at *www.natresnet.org*.

CHAPTER 24
Cooperative Buying

Community-Supported Agriculture

24.1 Shopping Locally

At the beginning of the twentieth century, 25 percent of Americans were directly involved in agriculture. By the 1960s, that proportion had fallen to five percent, even though the population increased dramatically. The two trends have accelerated, and although the United States has more than 300 million residents, only 1.8 percent are employed as farmers.

According to the U.S. Bureau of Labor Statistics, nearly half the agricultural workforce is self-employed, and the total number of people working as farmers and fishers continues to decline. Meanwhile, farms that generate more than $250,000 in annual sales make up less than 10 percent of all farms, but supply three-quarters of all agricultural output. This means fewer Americans than ever have direct contact with the way their food is grown.

A new way of growing and buying produce emerged in the late 1980s; it borrowed aspects of community life and commerce from the past, and introduced updated practices. In these ways, it offered an alternative to highly centralized and mechanized agriculture, sometimes called factory farming. This new method became known as community-supported agriculture (CSA).

The United States Department of Agriculture (USDA) uses the following definition of CSA: ". . . a community of individuals who pledge support to a farm operation so that the farmland becomes, either legally or spiritually, the community's farm, with the growers and consumers providing mutual support and sharing the risks and benefits of food production. Members or shareholders of the

iStockPhoto

farm or garden pledge in advance to cover the anticipated costs of the farm operation and farmer's salary. In return, they receive shares in the farm's bounty throughout the growing season, as well as satisfaction gained from reconnecting to the land. Members also share in risks, including poor harvest due to unfavorable weather or pests."

24.2 Origins of CSA

In some ways, having a community support a farm is as old as agriculture itself—but a few things make the current models different. For one thing, buying directly from a farmer is now the exception for most consumers, rather than the norm. For another, buyers are motivated by desires for fresh produce, high quality, local availability, and, for some, a connection with the land. This is directly at odds with customers driven only by price.

As it has become known, CSA began in Europe in the 1960s, and in Japan before that. The person most often credited with introducing it to the United States is Jan VanderTuin of Switzerland. He inspired the development of two CSA projects in the 1980s—one in Massachusetts and one in New Hampshire. The idea caught on quickly, and community groups across the country founded their own CSAs. Today, every state has some kind of CSA; the latest lists have more than 1500 farms.

Community-supported agriculture of North America at the University of Massachusetts—Extension traces the origins of the CSA concept to a group of women in Japan who were concerned about rising food imports and declining rates of farm employment. They established a direct buying system with local farms with the idea of "putting the farmers' face on food."

24.3   Philosophical Roots

Early leaders in the development of CSAs in the United States had several goals:

- Create community connections between city dwellers and rural residents.
- Encourage the development of sustainable growing practices.
- Bring control over food production back to the local level.
- Reduce or eliminate packaging and distribution costs.
- Develop strong local economies.

The movement also appealed to people for its rural training and employment opportunities, emphasis on local ownership of farmland, assurance of stable incomes for farmers, and independence from the control of remote commercial interests.

24.4   Subscription and Shareholder Models

Most CSAs use one of two models: subscription or shareholder.

A subscription CSA depends on the farmer to organize and manage the structure. In most cases, subscribers are not required to participate in the farm work. One variation on this model is a cooperative, in which two or more farmers collaborate to provide a greater variety of goods. Most CSAs are based on this model.

In a shareholder CSA, the customers form a buying group and seek out a farmer or group of farmers to grow the produce. The group may find an existing farm to hire, or it may decide to buy, lease, or rent farmland and take on a farmer as an employee. Members of the buying group may also work on the farm, but most will simply pay for their shares.

24.5   How a CSA Works

At the beginning of a year, the leaders of each CSA estimate their production costs for the growing season ahead. This includes everything from the tools that turn over the first row of soil to the containers that carry the last harvest to subscribers. The buying group divides that budget according to the volume each person or family has committed to buy. In the simplest structures, each individual pays the same amount of money and receives the same amount of produce. In practice, variations are almost inevitable.

As the growing season proceeds, each shareholder receives a portion of the regular harvests. The kind and variety of produce changes as different crops mature. In the most common arrangements, shares are bundled each week, intended to supply the vegetables a family of four will need until the next share is available. In general, shareholders pick up their own produce, or designate one person to pick up from the farm and deliver to others in the buying group. Depending on the number and size of farmers and buyers involved, the shares can range from vegetables to eggs, dairy products, fruit, honey, herbs, or flowers—even meat.

Some members of the buying group pay their full year's share at once; this helps provide the money needed for seeds, labor and other needs early in the season. Other members may pay for each share as it arrives; this makes CSA produce available to people who may not have enough money to buy their produce months in advance.

Not all CSA products are organic, and because the customers have already agreed to buy the produce, the farmer has no market incentive to pursue organic certification. But organic growing methods are consistent with CSA practices, so most of the produce does meet the commonly held definition of organic. Crops are often rotated during the year to provide variety in the weekly shares, and from year to year to improve the vitality of the soil. All crops depend on the local climate, which determines the length of the growing season.

24.6   Growing Variety

As CSA has become more popular, more innovations have appeared. Some CSAs have become land stewardship projects, others have begun offering home delivery, and still others have created apprenticeships, where the next generation of farmers will learn the trade. Each CSA has its own characteristics, which reflect the values of the people who participate.

Resources

To learn more about community-supported agriculture, and to find a CSA in your area, visit LocalHarvest at *www.localharvest.org*, the Land Stewardship Project at *www.landstewardshipproject.org* and the National Sustainable Agriculture Information Service (ATTRA) at *ww2.attra.ncat.org*.

CHAPTER 25

Agricultural Independence

Growing your own food

Container Gardening

25.1 Taking Charge

One way to make sure you have a consistent supply of fresh vegetables is to grow your own. If you don't have a garden plot, or even a yard, you can still raise several crops. The key is to plant them in containers rather than in the ground. This gives you a much greater choice of location; you can even move the containers as needed. If you have just a little room on a porch, patio, balcony, doorstep, or windowsill, you can start a container garden.

Even people with regular gardens enjoy container gardening because it lets them keep favorite vegetables and herbs close to the kitchen. It also keeps some plants out of the reach of pests, and lets you use exactly the right kind of soil for each crop.

25.2 Choosing a Container

Pretty much any vessel that can hold soil and allow excess water to drain out the bottom can be used for container gardening. It doesn't even need to be completely waterproof—but think twice before using materials that corrode, like metal cans, or rot, like woven wood baskets. Choose containers that are easy to handle, with appearances you like, or at least don't mind too much. There's no point in making a pleasant project less enjoyable.

iStockPhoto

The size of container you choose will depend on the size of the plant you hope to grow in it. Thin, vertical crops, such as parsley and other herbs, don't need too much room underneath for their roots. A container 6 to 10 inches across will hold them comfortably. The same goes for green onions. Larger vegetables, such as tomatoes and peppers, need more room. The most practical option for them is a 5-gallon bucket. It's also one of the least attractive, so you may want to find ways to dress it up.

In all containers of all sizes, make sure you provide a way for the soil to shed excess water. Drill or punch holes around the sides near the bottom for drainage, and put an inch or two of coarse gravel in the bottom before you add the soil.

25.2   Know What to Grow

Your choices for vegetables and herbs are limited only by your ambition, but you'll probably want to start with crops that are known to do well in container gardens. Many varieties of the following vegetables grow well in containers, even indoors:

- Cucumbers
- Eggplant
- Green Beans
- Green onions
- Leaf lettuce
- Peppers
- Radishes
- Squash
- Tomatoes

25.3   Seeding and Transplanting

Before you plant vegetables or herbs in containers, make sure they are germinating. Use temporary growing places, such as peat pots, egg cartons, or plastic trays intended for starting seeds. Cover each seed with a ¼-inch to a ½-inch of soil or other growing medium. Until they are ready to transplant, only use tap water. Keep the newly planted seeds in a warm, sunny place for 1 to 2 months. When each seedling has developed two or three full leaves, it should be ready for transplanting.

Whether you start from seeds or buy seedlings from a nursery, transfer them carefully to the container to avoid damaging the roots. Start using the growing solution the same day you transplant.

25.4   Growing Conditions

Soil

Once you choose a crop and prepare a container, you can start putting the parts together. The medium you select may not be soil, strictly speaking; mixtures that include sawdust, wood chips, peat moss, vermiculite, or perlite are often recommended for use in container gardens; check with a local garden center or a college agricultural extension service for recommendations. Of course, you can use soil, too. Just make sure it is lightweight and porous, and does not contain weed seeds or common diseases that affect plant life.

Another option is to make your own growing medium. Mix a bushel of peat moss with a bushel of horticultural grade vermiculite, then blend in the following minerals: 1 cup of garden fertilizer (6-12-12 or 5-10-10), 10 tablespoons of limestone, and 5 tablespoons of super-phosphate (0-20-0). Spray a mist of water over the mixture to keep the dust down, and stir it well. It will be suitable for transplanting or for use with new seeds.

Light

Most garden vegetables grow well in direct sunlight; remember, they are normally planted in the open. But fruit-bearing plants need more light than root vegetables, which need more than leafy ones. So, if you're growing cucumbers, tomatoes, peppers, or eggplant, put your containers in the brightest spots available. Give the next brightest spots to onions, beets, turnips, or radishes, and save the shadier places for lettuce, spinach, cabbage, and parsley.

If your space is limited, or if you only get sunlight for short periods in the places available, use containers with handles and move them during the day to get the light they need. This requires a serious commitment to gardening—but you may find, as many gardeners do, that it's just one more chore, and not a very big one.

Water

No matter where a plant grows, it needs water. But too much water in the soil can drown a plant from beneath. The soil (or other growing medium) has to drain so that the plant can draw oxygen through its roots. So water the containers every day or two, but watch out for saturation. And try not to water the leaves, because plant diseases can take hold more easily if the part of a plant above the surface is wet.

Fertilizer

The best time to add fertilizer is when you water. This makes things easier to remember, because plants that need water more often may also benefit from more frequent fertilization. Plants that are inactive will need less of both. Just keep in mind that fertilizer does nothing for the plant above the surface, so pour a solution of fertilizer and water directly into the soil.

If you buy a prepared fertilizing solution, follow the manufacturer's directions. To make your own concentrate, start with a gallon of warm tap water and add 2 cups of 10-20-10, 12-24-12, or 8-16-8. To use the concentrate, add two tablespoons to a gallon of water, and use this dilute solution for watering your plants. One gallon of concentrate will provide enough fertilizer for 256 gallons of growing solution.

Once a week, skip the fertilizer and use tap water only. This will help remove leftover nutrients the plant has not consumed. And once every month or so, use a nutrient solution with minor elements in place of your regular fertilizer. Buy a water-soluble compound that contains boron, iron, zinc, and manganese, and follow the package directions.

25.5   Plant Health

If you start with a clean planting medium and seeds or healthy transplants, you should not have any problems with insects or diseases. But keep an eye out for evidence of pests that feed on the leaves or fruit of your plants. If you are sure the plant has a disease or is infested, use only fungicides and insecticides that are approved by the EPA as safe for use on food.

Resources

The success of any container garden will also depend on the climate in which you live, and the vegetables and herbs you want to raise. To find out which combinations work best in your area, contact a local garden center or garden club, or the agricultural extension service of a nearby college.

CHAPTER 26

Eliminating Waste

Turning garbage into good soil

Composting Biodegradables

26.1   Organic Alchemy

The plant waste we generate, from food scraps to yard clippings, makes up nearly a quarter of the material dumped in landfills. Most of it can be composted instead, which both keeps biodegradable material out of the waste stream and generates valuable soil for use in lawns and gardens. Even if it didn't bring other benefits as well, those two things would make it worthwhile.

Composting is mostly passive. When you leave plant matter in a pile, it rots. Trouble is, it smells bad and isn't very useful. With just a few deliberate actions, though, you can turn this natural process into a productive one. Keep a balance of ingredients and stir up the pile every once in a while. There's more to it, but this chapter can help you get started.

26.2   The Composition of Decomposition

It starts as a heap of recognizable materials—apple cores, grass cuttings, melon rinds, nut shells, and so on—and ends up a stable source full of nutrients, ready to help another cycle of plant life grow. In between those two states, compost is a combination of life and death.

In nature, when dying plants fall to the ground, they begin to decay. The nutrients in the plant become food for other plants, animals, and microorganisms. Those not absorbed by decomposition return to the soil.

Intentional composting follows the same path, but with some important differences. First, the plant material is contained, so it can

iStockPhoto

be moved once it is finished decomposing. And because it's confined in a limited space, it becomes very hot—which kills organisms that might survive natural decomposition. The result is clean compost, free from weed seeds and pathogens that could affect the plants growing in it.

Compost can include any biodegradable material from food scraps to yard waste to manure. The speed with which it breaks down depends on the ratio of those ingredients, and on the presence of other materials that provide additional mass. Once this combination has finished decomposing, stabilizing, and curing, it is considered mature compost. The main ingredient in mature compost is humus, a dark and earthy substance that feels and smells like soil.

26.3   Ingredients for Success

The following biodegradable materials are suitable for use in compost, according to recommendation from the U.S. Environmental Protection Agency (EPA):

- Animal manure
- Cardboard rolls
- Cotton rags
- Dryer and vacuum cleaner lint
- Eggshells
- Fireplace ashes
- Fruits and vegetables

- Clean paper
- Grass clippings
- Hair and fur
- Hay and straw
- Houseplants
- Leaves
- Nut shells

- Coffee grounds and filters
- Sawdust
- Shredded newspaper
- Tea bags
- Wood chips
- Wool rags
- Yard trimmings

26.4   Items to Avoid

Some materials may seem appropriate for composting, but experience has shown that they do not help produce usable mature compost. Some release substances that can harm plants grown in the resulting compost, while others can impede the process of decomposition. Others create unpleasant odors, attract unwanted animals, or just plain don't break down very well. They include:

- Black walnut leaves or twigs
- Coal or charcoal ash
- Dairy products
- Diseased or infested plants

- Fats, grease, lard, or oils
- Meat or fish bones and scraps
- Pet wastes
- Yard clippings with pesticides

26.5   How Composting Works: Five Parts

Nutrients

Successful composting requires two kinds of ingredients: green and brown. That doesn't refer to their actual colors, but to the two complementary purposes they serve. Green materials include food scraps, yard waste, and manure, which are full of nitrogen; brown materials include dry leaves and wood chips, which have little nitrogen but lots of carbon. The biodegradable material you add to a compost pile will vary according to your circumstances, but you will quickly learn what mixture works best if you pay attention.

Particles

The same way your stomach digests food more easily after you chew it, a compost pile will break down organic matter into its component parts more quickly if they are chopped, shredded, ground, or otherwise reduced in size. It translates to more surface area on which the natural processes of decomposition can take place. Smaller particles can also help keep the temperature of the compost pile consistent. The only drawback to finely broken down material is that it might prevent air flow that brings oxygen, another important ingredient.

Water

The microorganisms that break down biodegradable material need water. It serves as a medium for the nutrients, helping them get from source to destination in every step of the process. If the ingredients in your compost pile already contain large amounts of moisture, you will probably not need to add any. Closed composting containers retain moisture better than those that sit in the open air.

Oxygen

Decomposition can be both aerobic and anaerobic. Most plant matter, though, breaks down more quickly and completely when oxygen is present. The challenge is to aerate the pile without letting it get too dry. Methods include turning over the pile with a shovel or pitchfork, turning a barrel on a central axis, and adding shredded newsprint to create air spaces within the compost.

Temperature

Most of the work in composting is done by microorganisms, so the idea is to create the conditions under which they work best. One critical ingredient is temperature, which can slow or stop the progress of these microorganisms if it's either too high or too low. At lower temperatures, the components do not break down fully, and simply rot. At higher temperatures, the microorganisms themselves may die.

While the surrounding air will affect the overall temperature of any compost pile, the best way to manage it is by ensuring the right mix of nutrients, particle size, water, and oxygen. To save yourself the frustration of unsuccessful compost, get advice from a friend or neighbor who has practiced composting for at least a few years.

26.6 Benefits of Using Compost

You can use compost as mulch, to amend poor soil, or in place of prepared fertilizer. Added to soil, compost improves texture and aeration. It helps sandy soils retain water, and loosens the structure of soils with high concentrations of clay. Compost increases the fertility of soil and encourages healthy root growth. The nutrients that remain in compost after it has cured will add nitrogen, potassium, and phosphorus to the soil.

Compost also provides these benefits:

- Helps remediate contaminated soil
- Reduces or eliminates organic and nonorganic contaminants
- Treats semivolatile and volatile organic compounds
- Helps keep heavy metals out of groundwater
- Helps reduce the risks of plant diseases and pests
- Reduces methane that would be generated by landfill
- Prevents erosion and silting on sloped ground
- Reduces the need for water, fertilizers, and pesticides

26.7 Composting Methods

The methods and tools required for composting vary as widely as the ingredients that go into it. Some of the most common include heaps, holding units, turning units, pit trenching, and sheet composting. Here is a very brief description of each:

Heaps are just static mounds of biodegradable material, usually about three feet high by five feet around. The shape naturally retains water, but the only way to aerate this kind of compost is to turn it over using hand tools.

Holding units are simply bins that hold compostable material from start to finish. They can even be used indoors, because the right balance of ingredients will not smell bad. These work slowly, but may be the best choice for people with limited space.

Turning units allow the user to tip or roll the container to improve aeration. This helps speed up the process of decomposition, and raises the temperature to help destroy weed seeds and disease organisms. They generally take up more room than holding units.

Pit trenching is just what it sounds like—a hole in the ground, filled with composting material and covered with a thick layer of soil. This method requires the least maintenance of all, but takes anywhere from a month to a year to complete.

Sheet composting uses a similarly passive means of encouraging materials to decompose. The difference is that instead of burying them in a single hole, you spread them out over a large area, usually a garden, and turn them under. This method offers the least control over nutrient balance.

Resources

To learn more about composting, contact your local garden center and visit the Web sites *www.compostguide.com, www.composting101.com,* and *www.epa.gov/msw/compost.htm.*

CHAPTER 27

Reducing Waste

Using Graywater

27.1 Saving Two Ways

A hundred years ago, clean water for drinking and washing was uncertain in all but the most settled areas; now we take it for granted everywhere. But at current rates of consumption, we use 500 billion gallons of fresh water every day in the United States. The problem with our water supply is no longer whether it's safe, but whether it will continue to be available.

You may remember from Chap. 13 that recent innovations in the design of plumbing fixtures have made it possible for homeowners to reduce their consumption of fresh water at little expense. Another method of conservation is a little more complicated, but it holds great promise: the use of graywater.

27.2 A Definition of Graywater

As it is commonly understood, graywater is used water from sinks, showers, bathtubs, and washing machines. It does not include water from dishwashers, or from any fixture that washes away human waste. Usually, that just means toilets, but it can include washing machines that are used to clean soiled diapers. For lack of a better term, this kind of waste is known as blackwater. Whether kitchen sinks produce graywater depends on the definition currently held by your local building codes; if in doubt, assume it's not safe to use.

In most households, more than half of the water used results in graywater waste. If that much can be diverted for landscaping and

iStockPhoto

other uses, it can represent large savings over using additional fresh water for the same purposes.

27.3   Basic Requirements

Unlike low-flow fixtures, which only require you to remove an existing plumbing part and thread another in its place, graywater collection depends on a completely new, separate set of drain lines. To divide your blackwater drains from those that carry graywater, you will need to undertake extensive plumbing work. If you're not confident in your abilities as a plumber, consider hiring a professional contractor. And whether you hire out the work or do it yourself, make sure you check all building code rules and get approval from an inspector before you begin.

New houses are much easier to fit for graywater collection, because they don't have drain lines in place yet. Everything starts from scratch. In houses that have already been completed, retrofitting is possible within limits. An older house with a concrete basement floor is especially difficult to retrofit. In such a case, only the washing machine could be fitted for graywater collection.

27.4   How Graywater Is Collected and Used

The simplest form of graywater reuse is emptying the remains from a pet's water dish into a potted plant before you refill the dish. If you think about it, an integrated whole-house graywater collection system connected to an automated landscape watering setup is really doing the same thing, just on a larger scale.

Determined recyclers have developed ways to collect graywater from bathtubs and washing machines, but these methods involve still more work. Carrying a gallon bucket outdoors and pouring it on the garden is little effort compared to draining a tub or washer and hauling the contents somewhere else by hand. Even the people who practice this method call it tedious and backbreaking.

If you're like most people, collecting the waste water from air conditioners, dehumidifiers, and other appliances seems like more trouble than it's worth. So, if graywater collection makes economic and ecological sense, you'll probably want a passive system that doesn't require constant attention.

In a system like that, new drain lines are installed, and the waste water from sinks, showers, bathtubs, and washing machines runs into a storage tank. Water from the tank can be pumped out by hand or with a mechanical pump. Because most health and safety rules only permit graywater to be used for landscape irrigation, it is generally routed to a drip-watering system. Spraying is usually not allowed. Professionally installed systems can include automatic valves that redirect the flow to the sewer if the storage tank is full.

27.5 Economic and Environmental Benefits

In parts of the country that regularly receive plenty of rainwater, such as the Pacific Northwest, graywater collection may not make as much economic sense. It still reduces the amount of fresh water consumed and waste water that flows into municipal treatment facilities. But in dry places, like the Southwest, water rationing is already a fact of life, even as landscaping represents a higher proportion of residential water use. Where water is scarce, saving water is more valuable.

By some estimates, the average American household uses up to 15 percent of its fresh water consumption to sustain plants. Nationwide, that could add up to savings of more than 6 billion gallons of water every day. Add the water used for landscaping, factor in the savings in processing of both fresh water and waste water, and the potential difference is significant.

27.6 Using Graywater—or Not

The Arizona Department of Environmental Quality provides helpful information on collecting and using graywater. Advice from the department includes the following tips:

Do:

- Filter your graywater with something as simple as a stocking to trap hair and lint.
- Frequently check your plants for evidence of overwatering or damage from organic material in graywater.
- Use graywater only for flood or drip irrigation.
- Divert graywater to your sewer or septic system if you are laundering diapers or dyeing clothes.
- Structure your irrigation system so it doesn't waste water by letting it percolate beyond the root zone.
- Use PVC or ABS piping.

Don't:

- Drink your graywater.
- Reuse water that contains hazardous chemicals from photo labs, car parts, or oily rags.
- Allow your graywater to pond because it can increase health risks and provide breeding grounds for mosquitoes.
- Reuse graywater for spray irrigation.
- Irrigate root or leaf crops (such as carrots or lettuce) that you'll eat.
- Reuse graywater if family members suffer from infectious diseases, such as diarrhea, hepatitis, or internal parasites.

Resources

To learn more about graywater collection and use, contact your local building inspections department. The staff there will have the most current information that applies to your location. Other useful resources include *www.epa.gov/WaterSense, www.graywater.net,* and *www.wateruseitwisely.com.*

CHAPTER 28

Washing Up

Household cleaning made greener

Making Your Own Cleaning Products

28.1 Cleaning House

Every household gets dirty, and every household needs cleaning. But not everyone needs to buy and use all the cleaning products that are available. We have simply become accustomed to buying separate cleaners for each of a dozen parts of the house.

Some specialized cleaning products serve unique purposes. For example, dishwashing detergent does an excellent job at one task; no other kind of soap does better in its place, and it does not work nearly as well for any other job. On the other hand, some cleaning products are specialized to the point that one container may never be completely used.

While some developments in consumer products have definitely brought better results, most of the innovations touted by companies that sell cleaning products are related to packaging and fragrance. If those are not important to you, consider making your own.

28.2 Health, Safety, and Comfort

Commercial cleaners are relatively inexpensive and they work well, so you may not feel that making your own cleaning products is a good use of time. But you may have other reasons to do it yourself. Store-bought cleaners may have strong and unpleasant ingredients or added scents, or you may be concerned about allergic and asthmatic reactions to them. People with bronchitis or sinusitis may be especially sensitive.

iStockPhoto

Whether you feel that commercially made cleaners pose health risks, or you just prefer knowing what goes into the products you use, you should find the formulas listed below easy to follow, and the results should match the kind of clean you have come to expect.

Just a note of caution: if you or someone in your household is very sensitive to chemical odors or vapors, be just as careful with these products as you would be with chemically based, commercially produced cleaners.

28.3 The Basic Ingredients

Baking soda: Use as a cleaner, deodorizer, scouring powder, and water softener.

Borax (sodium borate): Use as a cleaner, deodorizer, disinfectant, and water softener; also can be used for cleaning finished surfaces such as wallpaper, painted walls, and floors.

Citrus solvent: Use for cleaning oily items such as grease stains, adhesives, and paint brushes.

Corn starch: Use for cleaning windows, shampooing rugs and carpets, and polishing furniture.

Isopropyl alcohol: Use as a disinfectant; as an alternative, use ethanol or an alcohol-water solution.

Soap: Use any soap that does not contain fragrances, phosphorus, or petroleum products for all types of cleaning tasks.

Trisodium phosphate (TSP): Use for clearing drains, removing old paint, and other jobs that would otherwise require caustic chemicals. Handle TSP with caution.

Washing soda (sodium carbonate decahydrate): Use as a grease cutter, stain remover, water softener, or surface cleaner; just don't use it on aluminum.

White vinegar: Use for removing most grease, mildew, odors, stains, and waxes.

28.4   Recipes for Special Cleaning Jobs

The ingredients listed above are inexpensive, easy to find, and safe for use within limits. For example, TSP is safer than some equivalents, but it's still poisonous to ingest. You can also combine these ingredients to make specialized cleaners, using the following formulas. Be sure you test each product on a small area before you use it on an entire surface or object.

In the Kitchen

Dish soap: Use unscented liquid soap; add a few drops of vinegar for greasy dishes.

Dishwasher detergent: Combine equal parts of borax and washing soda; for hard water, increase the proportion of washing soda.

Drain cleaner for metal drains: Pour ½ cup baking soda down the drain; follow it with ½ cup of white vinegar. Wait 15 minutes, then pour boiling water down the drain to clear it. Note: do not use this method if you have already tried using a chemical drain cleaner.

Oven cleaner: Dampen the inside of the oven with a sponge and warm water. Make a paste of ¾ cup baking soda, ¼ cup salt and ¼ cup water; apply it to the oven walls with a sponge or damp cloth. Let sit overnight. Remove it with a spatula and wipe the surfaces clean, rubbing gently with fine steel wool if necessary.

Scouring powder: Use baking soda on a wet sponge to scrub surfaces without scratching.

In the Bathroom

Disinfectant: Combine two teaspoons of borax and four tablespoons of vinegar with three cups of hot water. Apply with a dampened cloth or a manual pump spray bottle.

Mold remover: Combine one part hydrogen peroxide (three percent) with two parts water in a spray bottle and spray on areas with mold. Let stand 1 hour, then rinse.

Toilet cleaner: Combine two parts borax with one part lemon juice and use with a scrub brush, or apply equal parts baking soda and white vinegar and scrub.

Tub and tile cleaner: Use baking soda on a wet sponge, or use a mixture of two parts borax and one part lemon juice.

Everywhere

All-purpose cleaner: Combine ½ cup white vinegar and ¼ cup baking soda or two teaspoons borax into ½ gallon of water. Use for light cleaning on smooth surfaces.

Air freshener: Combine baking soda and lemon juice in an open dish, and place it near the source of unwanted odors. Replace often.

Carpet cleaner: Combine equal parts white vinegar and water in a spray bottle. Spray the mixture directly on the stain and let sit for several minutes. Scrub with a brush or sponge and warm, soapy water. For heavy stains, combine equal parts salt, borax, and vinegar to form a paste. Rub it into the carpet and let it dry, then vacuum.

Furniture polish: Add a few drops of lemon oil to a ½ cup of warm water. Mix well and use to dampen a soft cotton cloth. Wipe furniture with damp cloth; follow with a dry, soft cloth.

Glass cleaner: Mix two teaspoons of white vinegar with one quart warm water; apply with a clean cloth and wipe dry.

Hardwood floor polish: Combine equal parts white vinegar and vegetable oil; rub in.

Resources

For many more formulas and detailed advice on using natural products for cleaning your house, try books like *Green Clean* by Linda Mason Hunter and Mikki Halpin, *The Naturally Clean Home* by Karyn Siegel-Maier, and *Organic Housekeeping* by Ellen Sandbeck.

CHAPTER 29

Second Chances

Reusing Containers

29.1 More Goods, Less Garbage

Most of the products you can buy arrive in containers—bottles, jars, cans, and boxes. These containers are made durable enough to withstand handling from the manufacturer to the distributor to the retailer to you. Once you have the product out of the package, it no longer has a useful purpose. It becomes more material for recycling—or worse, more bulk for the landfill. This chapter offers a few suggestions for reducing the amount of packaging you accumulate, or finding other uses for it.

29.2 A Packaging Paradox

Paper products, more than anything else, fill the country's waste stream. Paper packages make up a huge proportion of landfill mass. The packaging industry itself is concerned about the waste stream, and holds source reduction as one of its principles. Using the slogan "less waste in the first place," packagers work to minimize the material used to hold, protect, and present products to consumers.

Two problems make that goal difficult to pursue. One, consumers expect products to look perfect, and often choose one product over another because of the way it appears on the shelf. Two, packaging is often used in attempts to increase the perceived value of the contents. Clearly, the packaging industry can do better—but so can consumers.

iStockPhoto

Because almost every product available has to be conveyed somehow, no one can do away with packaging as an integral part of commerce. But you can reduce the amount of packaging included in what you buy.

29.3   Alternative Buying Tactics

While Shopping

About 20 years ago, few grocery stores offered foods in bulk; today, it's common practice. To reduce the amount of new packaging you consume, you can either use the thin but strong plastic bags provided by the grocery store, or you can save packaging waste entirely by bringing your own containers. Just remember to have one still empty when you check out, so the cashier will know how much weight to deduct from the total.

When you shop for groceries or other goods, take your own bags. Canvas shopping totes are light and easy to carry when they're empty, and a lot sturdier when they're full. Buy a few, if you don't already have some, and make a habit of using them. They will last for years.

Another kind of buying in bulk requires more coordination, but the results can be worth the effort. Warehouse-style stores offer common products in quantities that are too great for most households, but would provide generous shares for multiple households that agree to split the cost. These products still have commercial packaging, but often far less of it.

At Home

You can replace disposable paper towels and napkins with cloth dish-towels and napkins, which are sturdier—and can be washed and reused. Think of cloth for cleaning chores, too. As convenient as pre-moistened wipes can be, they create three kinds of waste: they combine paper, plastic, and soap in a separate package that only duplicates what you probably have in the cupboard.

Ounce for ounce, some of the most expensive common household products are salon-style shampoo and conditioner. They may not rival designer-brand perfumes, but their packages make up a large part of their retail price. Check in your area for specialty stores that will refill your hair-care product packages with more of the same, or a generic equivalent that may provide the same quality of results.

Also, with so many electronic devices in the average household, reusable batteries are more important than ever. Years ago, they helped reduce the amount of heavy metals, especially mercury, sent to landfills. Manufacturers worked to create disposable batteries with less and less mercury; they succeeded well enough that extracting the mercury fraction is economically impractical. So now, rechargeable batteries mostly help keep disposable batteries themselves out of the waste stream.

At Work

Restaurants that cater to business customers place a high priority on serving you quickly. Often, that means using disposable dishes and utensils. When you visit these restaurants, take only as much of each item as you really intend to use. Better still, find eateries that reuse dishes. You may find the experience more pleasant in other ways.

Taking your own lunch to work will give you more choices over what you eat, and help you save money at the same time. If you do take your lunch to work, it's just as easy to use durable dishes and utensils. And bring your own mug. After all, coffee tastes better from ceramic than from polystyrene.

Other reusable products at work include ink and toner cartridges. Many services will pick up your empty cartridges and offer a discount on refilled ones. If your current printer, fax, or copy machine is under a service contract, this option may conflict with it. Another potential problem is that some toner cartridges include microchips to prevent the use of third-party products. While this tactic mostly serves to protect the sales of the manufacturer's toner, the real risk may be in violating the warranty. Do some research before you adopt this approach.

By the way, even though researchers from Hewlett-Packard and Microsoft claim the paperless office is an imaginary goal, you can still pursue it. In *The Myth of the Paperless Office*, Abigail J. Sellen and Richard H. R. Harper show how electronic communication did not cause a reduction in paper use, but instead corresponded with a 40 percent increase in it. Bear in mind, though, that they described old habits. You can create new ones if you want.

Resources

When the useful life of a product is truly at an end—meaning it can't be repaired—you can still find ecologically sound ways to pass it on. One is to offer it for sale through a community bulletin board. Electronic ones such as craigslist are popular, but the cork one with push-pins down at the grocery store still works, too. You can also make it available through a free exchange network; examples include free-market and freecycle. People who accept your cast-off goods usually intend to make practical use of them.

CHAPTER 30

Getting Outside

Using native plants and grasses

Xeriscaping

30.1   Reinventing Landscapes

Nearly 30 years ago, a task force of interested specialists met in Colorado to discuss options for creating landscape designs that conserve water. At that conference, the word "xeriscaping" was coined. Taken from the word roots for "dry" and "scene," the term was intended to describe attractive landscapes that require less water than others.

Both the word and the concept gained popularity. Since then, people throughout the Southwest— Colorado, New Mexico, Nevada, Arizona, and California—have found ways to create landscapes that satisfy their aesthetic wishes while fitting well with dry climates.

30.2   Perceptions and Misconceptions

The term xeriscaping has a clear meaning, but it can be taken too literally. Some homeowners believe it means that only the driest elements are used. They picture yards full of sand or gravel, punctuated with a cactus or two and a few large rocks. But those are only a few of the many plants and minerals that can make up a beautiful and true xeriscape.

Another assumption on the part of unfamiliar audiences is that only native plants can be used in xeriscaping. While plants that already survive and thrive in dry climates are better suited for deliberate use in settings intended to conserve water, some imported or "alien" plants work well, too.

iStockPhoto

30.3 Seven Principles of Xeriscaping

Xeriscape Colorado, a leading promoter of dry-climate landscape design, defines xeriscaping as "water conservation through creative landscaping." The organization lists seven principles for creating an effective xeriscape:

- Planning and design—for water conservation and beauty
- Improving the soil—using compost or manure as needed
- Efficient irrigation—with properly designed systems
- Reduced turf areas—manageable sizes and shapes, and appropriate grasses
- Selecting appropriate plant material—grouping plants of similar water needs together
- Mulching—such as woodchips to reduce evaporation and keep the soil cool
- Proper maintenance—mowing, weeding, pruning, and fertilizing properly

You may notice that none of these principles require a particular kind of plant life. Instead, they help designers and homeowners create and maintain landscapes that conserve water, provides attractive planting options, thrives with little fertilization, pruning or maintenance, and minimizes pest and disease problems.

30.4   Saving Resources, Saving Money

By some estimates, as much as half the water consumed by households in the western United States goes to lawns and landscaping. As the population continues to grow in places like Phoenix and Las Vegas, this kind of consumption can not be sustained. Already, water levels at Lake Mead, the reservoir created by Hoover Dam, is 80 feet below normal level. The water has dropped consistently for more than 20 years.

Landscape irrigation is less vital to survival than water for drinking, washing, bathing, and carrying away sewage. When water shortages become severe, municipalities in the Southwest issue watering bans. These are likely to become more frequent and last longer.

In some ways, xeriscaping is a return to normal for desert communities. The practice of keeping grass lawns in a dry climate was never easy because it contradicts nature. By anticipating a time when water will be less plentiful, homeowners and landscape architects who adopt water conservation practices are creating the future.

Followed carefully, the principles of xeriscaping can reduce landscape water use by 60 percent or more. By keeping water waste to a minimum and designing landscapes that only need a little water, homeowners can cut their water bills and maintenance costs, while increasing their property values by as much as 15 percent.

30.5   Planning a Xeriscape

Even if you're familiar with many of the plants available in your area, talk with a specialist when you start planning a xeriscape. A professional landscape architect, gardener, or horticulturist will probably have and share knowledge that can save you time and money—or inspire you to try combinations or designs you might never have considered by yourself.

Before you choose which plants and other elements to use, decide how you want the space to function. If it will be a high-traffic area, make sure you have somewhere for people to walk. If you want to create a secluded garden for peace and quiet, look for plants that provide coverage and shade while fitting comfortably in a dry setting.

You may be tempted to imitate the most attractive design you see, or one that looks as if it would fit a house like yours, but resist temptation. You will be the most satisfied if the design you choose suits the way you really live.

When you create a xeriscape design, part of the challenge is selecting plants that will grow into the space. Along with layout and planting times, you will need to think about soil preparation. Thoughtful planning will keep your needs for water, fertilizer, pesticide, and other inputs to a minimum. In turn, these material savings will reduce your expenses.

30.6   Preparing the Soil

To provide a stable, healthy environment for plants and grasses, soil should be able to absorb ½ inch or more of water per hour. When the ground is dry and hard packed, it may not absorb enough water to sustain root systems—but if the soil contains too much sand, water may not stay in it long enough for plants to benefit.

Before you start working with the soil, have it tested for nutrient balance and organic content. The results should help you decide what approach to take. If you're not sure how to proceed, ask a horticulturist from the agricultural extension service at the nearest college.

To adjust the absorbency of soil, you can add amending materials. Compost can remedy many soil problems and provide the qualities needed for successful planting. Clean, cured compost can also help prevent plant diseases and infestations.

Whatever amending substances you add to the soil, make sure you work them in well below the surface. Ideally, you should use a tiller with tines that reach at least 6 inches down. Blended, tilled soil is best for new plantings and early growth. Shrubs and trees have special requirements for planting; get clear instructions for each type you choose, and follow any directions carefully.

30.7   Irrigating the Site

Installing an irrigation system can be as simple or complicated as you choose. It can consist of little more than a drip hose and a valve handle, or it can incorporate moisture sensors, automatic timers, micro-emitters, and other technologies. Use these two criteria for selecting an irrigation system, and you should find the right set of options:

- Compare the one-time cost of automatic components and systems to the expected long-term cost of water.
- Balance your desire for convenience against your wishes to take an active part in maintaining the xeriscape.

One more thing: wait until you have finished amending the soil and creating locations for planting before you install the irrigation system. This allows you to concentrate fully on each task without risk of harming the other work you have done.

Resources

To learn more about xeriscaping, visit Xeriscape Colorado, a program of Colorado WaterWise Council, at *www.xeriscape.org*. Another good resource is the California Integrated Waste Management Board Web site at *www.ciwmb.ca.gov*. A popular and comprehensive book on the subject is *Xeriscape Handbook: A How-To Guide to Natural, Resource-Wise Gardening* by Gayle Weinstein.

Index

X